Heavy Metals in Soils and Plants

Heavy Metals in Soils and Plants

Editor

Pushpika Freitas

scitus
academics

Heavy Metals in Soils and Plants

Edited by **Pushpika Freitas**

ISBN: 978-1-68117-069-5

Library of Congress Control Number: 2015935467

www.scitusacademics.com
Box No. 4766, 616 Corporate Way,
Suite 2, Valley Cottage,
NY 10989

Printed in United States of America on Acid Free Paper ∞

Contents

Preface

Soil, one of the most important natural resources, is becoming degraded due to anthropogenic activities such as mining, agricultural activities, sewage sludge, fossil fuel combustion, metallurgical and chemical industries and electronics. Soil is a crucial component of rural and urban environments, and in both places land management is the key to soil quality. This series of technical notes examines the urban activities that cause soil degradation, and the management practices that protect the functions urban societies demand from soil. This technical note focuses on heavy metal soil contamination. Mining, manufacturing, and the use of synthetic products e.g. pesticides, paints, batteries, industrial waste, and land application of industrial or domestic sludge can result in heavy metal contamination of urban and agricultural soils. Heavy metals also occur naturally, but rarely at toxic levels. Potentially contaminated soils may occur at old landfill sites particularly those that accepted industrial wastes, old orchards that used insecticides containing arsenic as an active ingredient, fields that had past applications of waste water or municipal sludge, areas in or around mining waste piles and tailings, industrial areas where chemicals may have been dumped on the ground, or in areas downwind from industrial sites. Excess heavy metal accumulation in soils is toxic to humans and other animals. Exposure to heavy metals is normally chronic, due to food chain transfer. Acute poisoning from heavy metals is rare through ingestion or dermal contact, but is possible.

Editor

Chapter 1

Heavy Metal Polluted Soils: Effect on Plants and Bioremediation Methods

G. U. Chibuike[1] and S. C. Obiora[2]

[1]Department of Soil Science, University of Nigeria, Nsukka, Nigeria
[2]Department of Geology, University of Nigeria, Nsukka, Nigeria

ABSTRACT

Soils polluted with heavy metals have become common across the globe due to increase in geologic and anthropogenic activities. Plants growing on these soils show a reduction in growth, performance, and yield. Bioremediation is an effective method of treating heavy metal polluted soils. It is a widely accepted method that is mostly carried out in situ; hence it is suitable for the establishment/reestablishment of crops on treated soils. Microorganisms and plants employ different mechanisms for the bioremediation of polluted soils. Using plants for the treatment of polluted soils is a more common approach in the bioremediation of heavy metal polluted soils. Combining both microorganisms and plants is an approach to bioremediation that ensures a more efficient clean-up of heavy metal polluted soils. However, success of this approach largely depends on the species of organisms involved in the process.

INTRODUCTION

Although heavy metals are naturally present in the soil, geologic and anthropogenic activities increase the concentration of these elements to amounts that are harmful to both plants and animals. Some of these activities include mining and smelting of metals, burning of fossil fuels, use of fertilizers and pesticides in agriculture, production of batteries and other metal products in industries, sewage sludge, and municipal waste disposal [1–3].

Growth reduction as a result of changes in physiological and biochemical processes in plants growing on heavy metal polluted soils has been recorded [4–6]. Continued decline in plant growth reduces yield which eventually leads to food insecurity. Therefore, the remediation of heavy metal polluted soils cannot be overemphasized.

Various methods of remediating metal polluted soils exist; they range from physical and chemical methods to biological methods. Most physical and chemical methods (such as encapsulation, solidification, stabilization, electrokinetics, vitrification, vapour extraction, and soil washing and flushing) are expensive and do not make the soil suitable for plant growth [7]. Biological approach (bioremediation) on the other hand encourages the establishment/reestablishment of plants on polluted soils. It is an environmentally friendly approach because it is achieved via natural processes. Bioremediation is also an economical remediation technique compared with other remediation techniques. This paper discusses the nature and properties of soils polluted with heavy metals. Plant growth and performance on these soils were examined. Biological approaches employed for the remediation of heavy metal polluted soils were equally highlighted.

HEAVY METAL POLLUTED SOILS

Heavy metals are elements that exhibit metallic properties such as ductility, malleability, conductivity, cation stability, and ligand specificity. They are characterized by relatively high density and high relative atomic weight with an atomic number greater than 20 [2]. Some heavy metals such as Co, Cu, Fe, Mn, Mo, Ni, V, and Zn are required in minute quantities by organisms. However, excessive amounts of these elements can become harmful to organisms. Other heavy metals such as Pb, Cd, Hg, and As (a metalloid but generally referred to as a heavy metal) do not have any beneficial effect on organisms and are thus regarded as the "main threats" since they are very harmful to both plants and animals.

Metals exist either as separate entities or in combination with other soil components. These components may include exchangeable ions sorbed on the surfaces of inorganic solids, nonexchangeable ions and insoluble inorganic metal compounds such as carbonates and phosphates, soluble metal compound or free metal ions in the soil solution, metal complex of organic materials, and metals attached to silicate minerals [7]. Metals bound to silicate minerals represent the background soil metal concentration and they do not cause contamination/pollution problems compared with metals that exist as separate entities or those present in high concentration in the other 4 components [8].

Soil properties affect metal availability in diverse ways. Harter [9] reported that soil pH is the major factor affecting metal availability in soil. Availability of Cd and Zn to the roots ofThlaspi caerulescens decreased with increases in soil pH [10]. Organic matter and hydrous ferric oxide have been shown to decrease heavy metal availability through immobilization of these metals [11]. Significant positive correlations have also been recorded between heavy metals and some soil physical properties such as moisture content and water holding capacity [12].

Other factors that affect the metal availability in soil include the density and type of charge in soil colloids, the degree of complexation with ligands, and the soil's relative surface area [7, 13]. The large interface and specific surface areas provided by soil colloids help in controlling the concentration of heavy metals in natural soils. In addition, soluble concentrations of metals in polluted soils may be reduced by soil particles with high specific surface area, though this may be metal specific [7]. For instance, Mcbride and Martínez [14] reported that addition of amendment consisting of hydroxides with high reactive surface area decreased the solubility of As, Cd, Cu, Mo, and Pb while the solubility of Ni and Zn was not changed. Soil aeration, microbial activity, and mineral composition have also been shown to influence heavy metal availability in soils [15].

Conversely, heavy metals may modify soil properties especially soil biological properties [16]. Monitoring changes in soil microbiological and biochemical properties after contamination can be used to evaluate the intensity of soil pollution because these methods are more sensitive and results can be obtained at a faster rate compared with monitoring soil physical and chemical properties [17]. Heavy metals affect the number, diversity, and activities of soil microorganisms. The toxicity of these metals on microorganisms depends on a number of factors such as soil temperature, pH, clay minerals, organic matter, inorganic anions and cations, and chemical forms of the metal [16, 18, 19].

There are discrepancies in studies comparing the effect of heavy metals on soil biological properties. While some researchers have recorded negative effect of heavy metals on soil biological properties [16, 17, 20], others have reported no relationship between high heavy metal concentrations and some soil (micro)biological properties [21]. Some of the inconsistencies may arise because some of these studies were conducted under laboratory conditions using artificially contaminated soils while others were carried out using soils from areas that are actually polluted in the field. Regardless of the origin of the soils used in these experiments, the fact that the effect of heavy metals on soil biological properties needs to be studied in more detail in order to fully understand the effect of these metals on the soil ecosystem remains. Further, it is advisable to use a wide range of methods (such as microbial biomass, C and N mineralization, respiration, and enzymatic activities) when studying effect of metals on soil biological properties rather than focusing on a single method since results obtained from use of different methods would be more comprehensive and conclusive.

The presence of one heavy metal may affect the availability of another in the soil and hence plant. In other words, antagonistic and synergistic behaviours exist among heavy metals. Salgare and Acharekar

[22] reported that the inhibitory effect of Mn on the total amount of mineralized C was antagonized by the presence of Cd. Similarly, Cu and Zn as well as Ni and Cd have been reported to compete for the same membrane carriers in plants [23]. In contrast, Cu was reported to increase the toxicity of Zn in spring barley [24]. This implies that the interrelationship between heavy metals is quite complex; thus more research is needed in this area. Different species of the same metal may also interact with one another. Abedin et al. [25] reported that the presence of arsenite strongly suppressed the uptake of arsenate by rice plants growing on a polluted soil.

EFFECT OF HEAVY METAL POLLUTED SOIL ON PLANT GROWTH

The heavy metals that are available for plant uptake are those that are present as soluble components in the soil solution or those that are easily solubilized by root exudates [26]. Although plants require certain heavy metals for their growth and upkeep, excessive amounts of these metals can become toxic to plants. The ability of plants to accumulate essential metals equally enables them to acquire other nonessential metals [27]. As metals cannot be broken down, when concentrations within the plant exceed optimal levels, they adversely affect the plant both directly and indirectly.

Some of the direct toxic effects caused by high metal concentration include inhibition of cytoplasmic enzymes and damage to cell structures due to oxidative stress [28, 29]. An example of indirect toxic effect is the replacement of essential nutrients at cation exchange sites of plants [30]. Further, the negative influence heavy metals have on the growth and activities of soil microorganisms may also indirectly affect the growth of plants. For instance, a reduction in the number of beneficial soil microorganisms due to high metal concentration may lead to decrease in organic matter decomposition leading to a decline in soil nutrients. Enzyme activities useful for plant metabolism may also be hampered due to heavy metal interference with activities of soil microorganisms. These toxic effects (both direct and indirect) lead to a decline in plant growth which sometimes results in the death of plant [31].

The effect of heavy metal toxicity on the growth of plants varies according to the particular heavy metal involved in the process. Table 1 shows a summary of the toxic effects of specific metals on growth, biochemistry, and physiology of various plants. For metals such as Pb, Cd, Hg, and As which do not play any beneficial role in plant growth, adverse effects have been recorded at very low concentrations of these metals in the growth medium. Kibra [32] recorded significant reduction in height of rice plants growing on a soil contaminated with 1 mgHg/kg. Reduced tiller and panicle formation also occurred at this concentration of Hg in the soil. For Cd, reduction in shoot and root growth in wheat plants occurred when Cd in the soil solution was as low as 5 mg/L [33]. Most of the reduction in growth parameters of plants growing on polluted soils can be attributed to reduced photosynthetic activities, plant mineral nutrition, and reduced activity of some enzymes [34].

Table 1: Effect of heavy metal toxicity on plants

Heavy metal	Plant	Toxic effect on plant	Reference
As	Rice (Oryza sativa)	Reduction in seed germination; decrease in seedling height; reduced leaf area and dry matter production	[35, 36]
	Tomato (Lycopersicon esculentum)	Reduced fruit yield; decrease in leaf fresh weight	[37]
	Canola (Brassica napus)	Stunted growth; chlorosis; wilting	[38]
Cd	Wheat (Triticum sp.)	Reduction in seed germination; decrease in plant nutrient content; reduced shoot and root length	[33, 39]
	Garlic (Allium sativum)	Reduced shoot growth; Cd accumulation	[40]
	Maize (Zea mays)	Reduced shoot growth; inhibition of root growth	[41]

Co	Tomato (Lycopersicon esculentum)	Reduction in plant nutrient content	[42]
	Mung bean (Vigna radiata)	Reduction in antioxidant enzyme activities; decrease in plant sugar, starch, amino acids, and protein content	[43]
	Radish (Raphanus sativus)	Reduction in shoot length, root length, and total leaf area; decrease in chlorophyll content; reduction in plant nutrient content and antioxidant enzyme activity; decrease in plant sugar, amino acid, and protein content	[44]
Cr	Wheat (Triticum sp.)	Reduced shoot and root growth	[45, 46]
	Tomato (Lycopersicon esculentum)	Decrease in plant nutrient acquisition	[47, 48]
	Onion (Allium cepa)	Inhibition of germination process; reduction of plant biomass	[49]
Cu	Bean (Phaseolus vulgaris)	Accumulation of Cu in plant roots; root malformation and reduction	[50]
	Black bindweed (Polygonum convolvulus)	Plant mortality; reduced biomass and seed production	[51]
	Rhodes grass (Chloris gayana)	Root growth reduction	[52]

Hg	Rice (Oryza sativa)	Decrease in plant height; reduced tiller and panicle formation; yield reduction; bioaccumulation in shoot and root of seedlings	[32, 53]
	Tomato (Lycopersicon esculentum)	Reduction in germination percentage; reduced plant height; reduction in flowering and fruit weight; chlorosis	[54]
Mn	Broad bean (Vicia faba)	Mn accumulation shoot and root; reduction in shoot and root length; chlorosis	[55]
	Spearmint (Mentha spicata)	Decrease in chlorophyll a and carotenoid content; accumulation of Mn in plant roots	[56]
	Pea (Pisum sativum)	Reduction in chlorophylls a and b content; reduction in relative growth rate; reduced photosynthetic O2 evolution activity and photosystem II activity	[57]
	Tomato (Lycopersicon esculentum)	Slower plant growth; decrease in chlorophyll concentration	[58]

Ni	Pigeon pea (Cajanus cajan)	Decrease in chlorophyll content and stomatal conductance; decreased enzyme activity which affected Calvin cycle and CO2 fixation	[59]
	Rye grass (Lolium perenne)	Reduction in plant nutrient acquisition; decrease in shoot yield; chlorosis	[60]
	Wheat (Triticum sp.)	Reduction in plant nutrient acquisition	[61, 62]
	Rice (Oryza sativa)	Inhibition of root growth	[63]
Pb	Maize (Zea mays)	Reduction in germination percentage; suppressed growth; reduced plant biomass; decrease in plant protein content	[64]
	Portia tree (Thespesia populnea)	Reduction in number of leaves and leaf area; reduced plant height; decrease in plant biomass	[65]
	Oat (Avena sativa)	Inhibition of enzyme activity which affected CO2 fixation	[66]

Zn	Cluster bean (Cyamopsis tetragonoloba)	Reduction in germination percentage; reduced plant height and biomass; decrease in chlorophyll, carotenoid, sugar, starch, and amino acid content	[67]
	Pea (Pisum sativum)	Reduction in chlorophyll content; alteration in structure of chloroplast; reduction in photosystem II activity; reduced plant growth	[68]
	Rye grass (Lolium perenne)	Accumulation of Zn in plant leaves; growth reduction; decrease in plant nutrient content; reduced efficiency of photosynthetic energy conversion	[69]

For other metals which are beneficial to plants, "small" concentrations of these metals in the soil could actually improve plant growth and development. However, at higher concentrations of these metals, reductions in plant growth have been recorded. For instance, Jayakumar et al. [42] reported that, at 50 mgCo/kg, there was an increase in nutrient content of tomato plants compared with the control. Conversely, at 100 mgCo/kg to 250 mgCo/kg, reductions in plant nutrient content were recorded. Similarly, increase in plant growth, nutrient content, biochemical content, and antioxidant enzyme activities (catalase) was observed in radish and mung bean at 50 mgCo/kg soil concentration while reductions were recorded at 100 mgCo/kg to 250 mgCo/kg soil concentration [43, 44]. Improvements in growth and physiology of cluster beans have also been reported at Zn concentration of 25 mg/L of the soil solution. On the other hand, growth reduction and adverse effect on the plant's physiology started when the soil solution contained 50 mgZn/L [67].

It is worth mentioning that, in most real life situations (such as disposal of sewage sludge and metal mining wastes) where soil may be polluted with more than one heavy metal, both antagonistic and synergistic relationships between heavy metals may affect plant metal toxicity. Nicholls and Mal [70] reported that the combination of Pb and Cu at both high concentration (1000 mg/kg each) and low concentration (500 mg/kg) resulted in a rapid and complete death of the leaves and stem of Lythrum salicaria. The authors reported that there was no synergistic interaction between these heavy metals probably because the concentrations used in the experiment were too high for interactive relationship to be observed between the metals. Another study [71] examined the effect of 6 heavy metals (Cd, Cr, Co, Mn, and Pb) on the growth of maize. The result showed that the presence of these metals in soil reduced the growth and protein content of maize. The toxicity of these metals occurred in the following order: Cd > Co > Hg > Mn > Pb > Cr. It was also observed in this study that the combined effect of 2 or more heavy metals was only as harmful as the effect of the most toxic heavy metal. The researcher attributed this result to the antagonistic relationship which exists between heavy metals.

It is important to note that certain plants are able to tolerate high concentration of heavy metals in their environment. Baker [72] reported that these plants are able to tolerate these metals via 3 mechanisms, namely, (i) exclusion: restriction of metal transport and maintenance of a constant metal concentration in the shoot over a wide range of soil concentrations; (ii) inclusion: metal concentrations in the shoot reflecting those in the soil solution through a linear relationship; and (iii) bioaccumulation: accumulation of metals in the shoot and roots of plants at both low and high soil concentrations.

BIOREMEDIATION OF HEAVY METAL POLLUTED SOILS

Bioremediation is the use of organisms (microorganisms and/or plants) for the treatment of polluted soils. It is a widely accepted method of soil remediation because it is perceived to occur via natural processes. It is equally a cost effective method of soil remediation. Blaylock et al. [73] reported 50% to 65% saving when bioremediation was used for the treatment of 1 acre of Pb polluted soil compared with the case when a conventional method (excavation and landfill) was used for the same purpose. Although bioremediation is a nondisruptive method of soil remediation, it is usually time consuming and its use for the treatment of heavy metal polluted soils is sometimes affected by the climatic and geological conditions of the site to be remediated [74].

Heavy metals cannot be degraded during bioremediation but can only be transformed from one organic complex or oxidation state to another. Due to a change in their oxidation state, heavy metals can be transformed to become either less toxic, easily volatilized, more water soluble (and thus can be removed through leaching), less water soluble (which allows them to precipitate and become easily removed from the environment) or less bioavailable [75,76].

Bioremediation of heavy metals can be achieved via the use of microorganisms, plants, or the combination of both organisms.

Using Microbes for Remediation of Heavy Metal Polluted Soils

Several microorganisms especially bacteria (Bacillus subtilis, Pseudomonas putida, andEnterobacter cloacae) have been successfully used for the reduction of Cr (VI) to the less toxic Cr (III) [77–80]. B. subtilis has also been reported to reduce nonmetallic elements. For instance, Garbisu et al. [81] recorded that B. subtilis reduced the selenite to the less toxic elemental Se. Further, B. cereus and B. thuringiensis have been shown to increase extraction of Cd and Zn from Cd-rich soil and soil polluted with effluent from metal industry [82]. It is assumed that the production of siderophore (Fe complexing molecules) by bacteria may have facilitated the extraction of these metals from the soil; this is because heavy metals have been reported to simulate the production of siderophore and this consequently affects their bioavailability [83]. For instance, siderophore production by Azotobacter vinelandii was increased in the presence of Zn (II) [84]. Hence, heavy metals influence the activities of siderophore-producing bacteria which in turn increases mobility and extraction of these metals in soil.

Bioremediation can also occur indirectly via bioprecipitation by sulphate reducing bacteria (Desulfovibrio desulfuricans) which converts sulphate to hydrogen sulphate which subsequently reacts with heavy metals such as Cd and Zn to form insoluble forms of these metal sulphides [85].

Most of the above microbe assisted remediation is carried out ex situ. However, a very important in situ microbe assisted remediation is the microbial reduction of soluble mercuric ions Hg (II) to volatile metallic mercury and Hg (0) carried out by mercury resistant bacteria [86]. The reduced Hg (0) can easily volatilize out of the environment and subsequently be diluted in the atmosphere [87].

Genetic engineering can be adopted in microbe assisted remediation of heavy metal polluted soils. For instance, Valls et al. [88] reported that genetically engineered Ralstonia eutrophacan be used to sequester metals (such as Cd) in polluted soils. This is made possible by the introduction of metallothionein (cysteine rich metal binding protein) from mouse on the cell surface on this organism. Although the sequestered metals remain in the soil, they are made less bioavailable and hence less harmful. The controversies surrounding genetically modified organisms [89] and the fact that the heavy metal remains in the soil are major limitations to this approach to bioremediation.

Making the soil favourable for soil microbes is one strategy employed in bioremediation of polluted soils. This process known as biostimulation involves the addition of nutrients in the form of manure or other organic amendments which serve as C source for microorganisms present in the soil. The added nutrients increase the growth and activities of microorganisms involved in the remediation process and thus this increases the efficiency of bioremediation.

Although biostimulation is usually employed for the biodegradation of organic pollutants [90], it can equally be used for the remediation of heavy metal polluted soils. Since heavy metals cannot be biodegraded, biostimulation can indirectly enhance remediation of heavy metal polluted soil through alteration of soil pH. It is well known that the addition of organic materials reduces the pH of the soil [91]; this subsequently increases the solubility and hence bioavailability of heavy metals which can then be easily extracted from the soil [92].

Biochar is one organic material that is currently being exploited for its potential in the management of heavy metal polluted soils. Namgay et al. [93] recorded a reduction in the availability of heavy metals when the polluted soil was amended with biochar; this in turn reduced plant absorption of the metals. The ability of biochar to increase soil pH unlike most other organic amendments [94] may have increased sorption of these metals, thus reducing their bioavailability for plant uptake. It is important to note that, since the characteristics of biochar vary widely depending on its method of production and the feedstock used in its production, the effect different biochar amendments will have on the availability of heavy metals in soil will also differ. Further, more research is needed in order to understand the effect of biochar on soil microorganisms and how the interaction between biochar and soil microbes influences remediation of heavy metal polluted soils because such studies are rare in literature.

Using Plants for Remediation of Heavy Metal Polluted Soils

Phytoremediation is an aspect of bioremediation that uses plants for the treatment of polluted soils. It is suitable when the pollutants cover a wide area and when they are within the root zone of the plant [76]. Phytoremediation of heavy metal polluted soils can be achieved via different mechanisms. These mechanisms include phytoextraction, phytostabilization, and phytovolatilization.

Phytoextraction

This is the most common form of phytoremediation. It involves accumulation of heavy metals in the roots and shoots of phytoremediation plants. These plants are later harvested and incinerated. Plants used for phytoextraction usually possess the following characteristics: rapid growth rate, high biomass, extensive root system, and ability to tolerate high amounts of heavy metals. This ability to tolerate high concentration of heavy metals by these plants may lead to metal accumulation in the harvestable part; this may be problematic through contamination of the food chain [7].

There are two approaches to phytoextraction depending on the characteristics of the plants involved in the process. The first approach involves the use of natural hyperaccumulators, that is, plants with very high metal-accumulating ability, while the second approach involves the use of high biomass plants whose ability to accumulate metals is induced by the use of chelates, that is, soil amendments with metal mobilizing capacity [95].

Hyperaccumulators accumulate 10 to 500 times more metals than ordinary plant [96]; hence they are very suitable for phytoremediation. An important characteristic which makes hyperaccumulation possible is the tolerance of these plants to increasing concentrations of these metals (hypertolerance). This could be a result of exclusion of these metals from the plants or by compartmentalization of these metal ions; that is, the metals are retained in the vacuolar compartments or cell walls and thus do not have access to cellular sites where vital functions such as respiration and cell division take place [76, 96].

Generally, a plant can be called a hyperaccumulator if it meets the following criteria: (i) the concentration of metal in the shoot must be higher than 0.1% for Al, As, Co, Cr, Cu, Ni, and Se, higher than 0.01% for Cd, and higher than 1.0% for Zn [97]; (ii) the ratio of shoot to root concentration must be consistently higher than 1 [98]; this indicates the capability to transport metals from roots to shoot and the existence of hypertolerance ability [7]; (iii) the ratio of shoot to root concentration must be higher than 1; this indicates the degree of plant metal uptake [7, 98]. Reeves and Baker [99] reported some examples of plants which have the ability to accumulate large amounts of heavy metals and hence can be used in remediation studies. Some of these plants include Haumaniastrum robertii (Co hyperaccumulator); Aeollanthus subacaulis (Cu hyperaccumulator); Maytenus bureaviana(Mn hyperaccumulator); Minuartia verna and Agrostis tenuis (Pb hyperaccumulators);Dichapetalum gelonioides, Thlaspi tatrense, and Thlaspi caerulescens (Zn hyperaccumulators); Psycotria vanhermanni and Streptanthus polygaloides (Ni hyperaccumulators); Lecythis ollaria (Se hyperaccumulator). Pteris vittata is an example of a hyperaccumulator that can be used for the remediation of soils polluted with As [100]. Some plants have the ability to accumulate more than one metal. For instance, Yang et al. [101] observed that the Zn hyperaccumulator, Sedum alfredii, can equally hyperaccumulate Cd.

The possibility of contaminating the food chain through the use of hyperaccumulators is a major limitation in phytoextraction. However, many species of the Brassicaceae family which are known to be hyperaccumulators of heavy metals contain high amounts of thiocyanates which make them unpalatable to animals; thus this reduces the availability of these metals in the food chain [102].

Most hyperaccumulators are generally slow growers with low plant biomass; this reduces the efficiency of the remediation process [103]. Thus, in order to increase the efficiency of phytoextraction, plants with high growth rate as well as high biomass (e.g., maize, sorghum, and alfalfa) are sometimes used together with metal chelating substances for soil remediation exercise. It is important to note that some hyperaccumulators such as certain species within the Brassica genus (Brassica napus, Brassica juncea, and Brassica rapa) are fast growers with high biomass [104].

In most cases, plants absorb metals that are readily available in the soil solution. Although some metals are present in soluble forms for plant uptake, others occur as insoluble precipitate and are thus unavailable for plant uptake. Addition of chelating substances prevents precipitation and metal sorption via the formation of metal chelate complexes; this subsequently increases the bioavailability of these metals [7]. Further, the addition of chelates to the soil can transport more metals into the soil solution through the dissolution of precipitated compounds and desorption of sorbed species [13]. Certain chelates are also able to translocate heavy metal into the shoots of plants [73].

Marques et al. [7] documented examples of synthetic chelates which have successfully been used to extract heavy metals from polluted soils. Some of these chelates include EDTA (ethylenediaminetetraacetic acid), EDDS (SS-ethylenediamine disuccinic acid), CDTA (trans-1,2-diaminocyclohexane-N,N,N′,N′-tetraacetic acid), EDDHA (ethylenediamine-di-o-hydroxyphenylacetic acid), DTPA (diethylenetriaminepentaacetic acid), and HEDTA (N-hydroxyethylenediaminetriacetic acid). EDTA is a synthetic chelate that is widely used not only because it is the least expensive compared with other synthetic chelates [105] but also because it has a high ability to successfully improve plant metal uptake [106–108]. Organic chelates such as citric acid and malic acid can also be used to improve phytoextraction of heavy metals from polluted soils [109].

One major disadvantage of using chelates in phytoextraction is the possible contamination of groundwater via leaching of these heavy metals [110]. This is because of the increased availability of heavy metals in the soil solution when these chelates are used. In addition, when chelates (especially synthetic chelates) are used in high concentrations, they can become toxic to plants and soil microbes

[106]. In general, solubility/availability of heavy metals for plant uptake and suitability of a site for phytoextraction are additional factors that should be considered (in addition to suitability of plants) before using phytoextraction for soil remediation [26].

Phytostabilization

Phytostabilization involves using plants to immobilize metals, thus reducing their bioavailability via erosion and leaching. It is mostly used when phytoextraction is not desirable or even possible [98]. Marques et al. [7] argued that this form of phytoremediation is best applied when the soil is so heavily polluted so that using plants for metal extraction would take a long time to be achieved and thus would not be adequate. Jadia and Fulekar [111] on the other hand showed that the growth of plants (used for phytostabilization) was adversely affected when the concentration of heavy metal in the soil was high.

Phytostabilization of heavy metals takes place as a result of precipitation, sorption, metal valence reduction, or complexation [29]. The efficiency of phytostabilization depends on the plant and soil amendment used. Plants help in stabilizing the soil through their root systems; thus, they prevent erosion. Plant root systems equally prevent leaching via reduction of water percolation through the soil. In addition, plants prevent man's direct contact with pollutants and they equally provide surfaces for metal precipitation and sorption [112].

Based on the above factors, it is important that appropriate plants are selected for phytostabilization of heavy metals. Plants used for phytostabilization should have the following characteristics: dense rooting system, ability to tolerate soil conditions, ease of establishment and maintenance under field conditions, rapid growth to provide adequate ground coverage, and longevity and ability to self-propagate.

Soil amendments used in phytostabilization help to inactivate heavy metals; thus, they prevent plant metal uptake and reduce biological activity [7]. Organic materials are mostly used as soil amendments in phytostabilization. Marques et al. [113] showed that Zn percolation through the soil reduced by 80% after application of manure or compost to polluted soils on which Solanum nigrum was grown.

Other amendments that can be used for phytostabilization include phosphates, lime, biosolids, and litter [114]. The best soil amendments are those that are easy to handle, safe to workers who apply them, easy to produce, and inexpensive and most importantly are not toxic to plants [113]. Most of the times, organic amendments are used because of their low cost and the other benefits they provide such as provision of nutrients for plant growth and improvement of soil physical properties [7].

In general, phytostabilization is very useful when rapid immobilization of heavy metals is needed to prevent groundwater pollution. However, because the pollutants remain in the soil, constant monitoring of the environment is required and this may become a problem.

Phytovolatilization

In this form of phytoremediation, plants are used to take up pollutants from the soil; these pollutants are transformed into volatile forms and are subsequently transpired into the atmosphere [115]. Phytovolatilization is mostly used for the remediation of soils polluted with Hg. The toxic form of Hg (mercuric ion) is transformed into the less toxic form (elemental Hg). The problem with this process is that the new product formed, that is, elemental Hg, may be redeposited into lakes and rivers after being recycled by precipitation; this in turn repeats the process of methyl-Hg production by anaerobic bacteria [115].

Raskin and Ensley [116] reported the absence of plant species with Hg hyperaccumulating properties. Therefore, genetic engineered plants are mostly used in phytovolatilization. Examples of transgenic plants which have been used for phytovolatilization of Hg polluted soils are Nicotiana tabacum, Arabidopsis thaliana, and Liriodendron tulipifera [117, 118]. These plants are usually genetically modified to include gene for mercuric reductase, that is, merA. Organomercurial lyase (merB) is another bacterial gene used for the detoxification of methyl-Hg. Both merA and merB can be inserted into plants used to detoxify methyl-Hg to elemental Hg [119]. Use of plants modified with merA and merB is not acceptable from a regulatory perspective [119]. However, plants altered with merB are more acceptable because the gene prevents the introduction of methyl-Hg into the food chain [120].

Phytovolatilization can also be employed for the remediation of soils polluted with Se [7]. This involves the assimilation of inorganic Se into organic selenoamino acids (selenocysteine and selenomethionine). Selenomethionine is further biomethylated to dimethylselenide which is lost in the atmosphere via volatilization [121]. Plants which have successfully been used for phytovolatilization of soils polluted with Se are Brassica juncea and Brassica napus[122].

Combining Plants and Microbes for the Remediation of Heavy Metal Polluted Soils

The combined use of both microorganisms and plants for the remediation of polluted soils results in a faster and more efficient clean-up of the polluted site [123]. Mycorrhizal fungi have been used in several remediation studies involving heavy metals and the results obtained show that mycorrhizae employ different mechanisms for the remediation of heavy metal polluted soils. For instance, while some studies have shown enhanced phytoextraction through the accumulation of heavy metals in plants [124–126], others reported enhanced phytostabilization through metal immobilization and a reduced metal concentration in plants [127, 128].

In general, the benefits derived from mycorrhizal associations—which range from increased nutrient and water acquisition to the provision of a stable soil for plant growth and increase in plant resistance to diseases [129–131]—are believed to aid the survival of plants growing in polluted soils and thus help in the vegetation/revegetation of remediated soils [132]. It is important to note that mycorrhiza does not always assist in the remediation of heavy metal polluted soils [133, 134] and this may be attributed to the species of mycorrhizal fungi and the concentration of heavy metals [7, 132]. Studies have also shown that activities of mycorrhizal fungi may be inhibited by heavy metals [135, 136]. In addition, Weissenhorn and Leyval [137] reported that certain species of mycorrhizal fungi (arbuscular mycorrhizal fungi) can be more sensitive to pollutants compared to plants.

Other microorganisms apart from mycorrhizal fungi have also been used in conjunction with plants for the remediation of heavy metal polluted soils. Most of these microbes are the plant growth-promoting rhizobacteria (PGPR) that are usually found in the rhizosphere. These PGPR stimulate plant growth via several mechanisms such as production of phytohormones and supply of nutrients [138], production of siderophores and other chelating agents [139], specific enzyme activity and N fixation [140], and reduction in ethylene production which encourages root growth [141].

In general, PGPR have been used in phytoremediation studies to reduce plant stress associated with heavy metal polluted soils [142]. Enhanced accumulation of heavy metals such as Cd and Ni by hyperaccumulators (Brassica juncea and Brassica napus) has been observed when the plants were inoculated with Bacillus sp. [143, 144]. On the other hand, Madhaiyan et al. [145] reported increased plant growth due to a reduction in the accumulation of Cd and Ni in the shoot and root tissues of tomato plant when it was inoculated with Methylobacterium oryzae and Burkholderia spp. Thus, this indicates that the mechanisms employed by PGPR in the phytoremediation of heavy metal polluted soils may be dependent on the species of PGRP and plant involved in the process. Although studies involving both the use of mycorrhizal fungi and PGPR are uncommon, Vivas et al. [146] reported that PGPR (Brevibacillus sp.) increased mycorrhizal efficiency which in turn decreased metal accumulation and increased the growth of white clover growing on a heavy metal (Zn) polluted soil.

CONCLUSIONS

Plants growing on heavy metal polluted soils show a reduction in growth due to changes in their physiological and biochemical activities. This is especially true when the heavy metal involved does not play any beneficial role towards the growth and development of plants. Bioremediation can be effectively used for the treatment of heavy metal polluted soil. It is most appropriate when the remediated site is used for crop production because it is a nondisruptive method of soil remediation. Using plants for bioremediation (phytoremediation) is a more common approach to bioremediation of heavy metal compared with the use of microorganisms. Plants employ different mechanisms in the remediation of heavy metal polluted soils. Phytoextraction is the most common method of phytoremediation used for treatment of heavy metal polluted soils. It ensures the complete removal of the pollutant. Combining both plants and microorganisms in bioremediation increases the efficiency of this method of remediation. Both mycorrhizal fungi and other PGPR have been successfully incorporated in various phytoremediation programmes. The success of the combined use of these organisms depends on the species of microbe and plants involved and to some extent on the concentration of the heavy metal in soil.

REFERENCES

1. B. J. Alloway, Heavy Metal in Soils, John Wiley & Sons, New York, NY, USA, 1990.
2. I. Raskin, P. B. A. N. Kumar, S. Dushenkov, and D. E. Salt, "Bioconcentration of heavy metals by plants," Current Opinion in Biotechnology, vol. 5, no. 3, pp. 285–290, 1994.
3. Z. Shen, X. Li, C. Wang, H. Chen, and H. Chua, "Lead phytoextraction from contaminated soil with high-biomass plant species," Journal of Environmental Quality, vol. 31, no. 6, pp. 1893–1900, 2002.
4. J. Chatterjee and C. Chatterjee, "Phytotoxicity of cobalt, chromium and copper in cauliflower," Environmental Pollution, vol. 109, no. 1, pp. 69–74, 2000.
5. I. Öncel, Y. Keleş, and A. S. Üstün, "Interactive effects of temperature and heavy metal stress on the growth and some biochemical compounds in wheat seedlings,"Environmental Pollution, vol. 107, no. 3, pp. 315–320, 2000.
6. S. Oancea, N. Foca, and A. Airinei, "Effects of heavy metals on plant growth and photosynthetic activity," Analele Ştiinţifice ale Universităţii "AL. I. CUZA1 IAŞI, Tomul I, s. Biofizică, Fizică medicală şi Fizica mediului, pp. 107–110, 2005.
7. A. P. G. C. Marques, A. O. S. S. Rangel, and P. M. L. Castro, "Remediationofheavymetalcontaminatedsoils:phytoremediation as a potentially promising clean-up technology," Critical Reviews in Environmental Science and Technology, vol. 39, no. 8, pp. 622–654, 2009.
8. L. Ramos, L. M. Hernandez, and M. J. Gonzalez, "Sequential fractionation of copper, lead, cadmium and zinc in soils from or near Donana National Park," Journal of Environmental Quality, vol. 23, no. 1, pp. 50–57, 1994.
9. R. D. Harter, "Effect of soil pH on adsorption of lead, copper, zinc, and nickel," Soil Science Society of America Journal, vol. 47, no. 1, pp. 47–51, 1983.

10. A. S. Wang, J. S. Angle, R. L. Chaney, T. A. Delorme, and R. D. Reeves, "Soil pH effects on uptake of Cd and Zn by Thlaspi caerulescens," Plant and Soil, vol. 281, no. 1-2, pp. 325–337, 2006.
11. L. Yi, Y. Hong, D. Wang, and Y. Zhu, "Determination of free heavy metal ion concentrations in soils around a cadmium rich zinc deposit," Geochemical Journal, vol. 41, no. 4, pp. 235–240, 2007.
12. M. S. Rakesh Sharma and N. S. Raju, "Correlation of heavy metal contamination with soil properties of industrial areas of Mysore, Karnataka, India by cluster analysis,"International Research Journal of Environment Sciences, vol. 2, no. 10, pp. 22–27, 2013.
13. W. A. Norvell, "Comparison of chelating agents as extractants for metals in diverse soil materials," Soil Science Society of America Journal, vol. 48, no. 6, pp. 1285–1292, 1984.
14. M. B. Mcbride and C. E. Martínez, "Copper phytotoxicity in a contaminated soil: remediation tests with adsorptive materials," Environmental Science and Technology, vol. 34, no. 20, pp. 4386–4391, 2000.
15. M. L. Magnuson, C. A. Kelty, and K. C. Kelty, "Trace metal loading on water-borne soil and dust particles characterized through the use of Split-flow thin-cell fractionation," Analytical Chemistry, vol. 73, no. 14, pp. 3492–3496, 2001.
16. M. Friedlová, "The influence of heavy metals on soil biological and chemical properties," Soil and Water Research, vol. 5, no. 1, pp. 21–27, 2010.
17. P. Nannipieri, L. Badalucco, L. Landi, and G. Pietramellara, "Measurement in assessing the risk of chemicals to the soil ecosystem," in Ecotoxicology: Responses, Biomarkers and Risk Assessment, J. T. Zelikoff, Ed., pp. 507–534, OECD Workshop, SOS Publ., Fair Haven, NY, USA, 1997.
18. E. Baath, "Effects of heavy metals in soil on microbial processes and populations (a review)," Water, Air, & Soil Pollution, vol. 47, no. 3-4, pp. 335–379, 1989.
19. K. E. Giller, E. Witter, and S. P. Mcgrath, "Toxicity of heavy metals to microorganisms and microbial processes in agricultural soils," Soil Biology and Biochemistry, vol. 30, no. 10-11, pp. 1389–1414, 1998.

20. M. Šmejkalova, O. Mikanova, and L. Borůvka, "Effects of heavy metal concentrations on biological activity of soils microorganisms," Plant, Soil and Environment, vol. 49, pp. 321–326, 2003.
21. S. Castaldi, F. A. Rutigliano, and A. Virzo de Santo, "Suitability of soil microbial parameters as indicators of heavy metal pollution," Water, Air, & Soil Pollution, vol. 158, no. 1, pp. 21–35, 2004.
22. S. A. Salgare and C. Acharekar, "Effect of industrial pollution on growth and content of certain weeds," Journal for Nature Conservation, vol. 4, pp. 1–6, 1992.
23. D. T. Clarkson and U. Luttge, "Mineral nutrition: divalent cations, transport and compartmentation," Progress in Botany, vol. 51, pp. 93–112, 1989.
24. Y. Luo and D. L. Rimmer, "Zinc-copper interaction affecting plant growth on a metal-contaminated soil," Environmental Pollution, vol. 88, no. 1, pp. 79–83, 1995.
25. M. J. Abedin, J. Feldmann, and A. A. Meharg, "Uptake kinetics of arsenic species in rice plants," Plant Physiology, vol. 128, no. 3, pp. 1120–1128, 2002.
26. M. J. Blaylock and J. W. Huang, "Phytoextraction of metals," in Phytoremediation of Toxic Metals: Using Plants to Clean up the Environment, I. Raskin and B. D. Ensley, Eds., pp. 53–70, Wiley, New York, NY, USA, 2000.
27. R. Djingova and I. Kuleff, "Instrumental techniques for trace analysis," in Trace Elements: Their Distribution and Effects in the Environment, J. P. Vernet, Ed., Elsevier, London, UK, 2000.
28. F. Assche and H. Clijsters, "Effects of metals on enzyme activity in plants," Plant, Cell and Environment, vol. 24, pp. 1–15, 1990.
29. C. D. Jadia and M. H. Fulekar, "Phytoremediation of heavy metals: recent techniques,"African Journal of Biotechnology, vol. 8, no. 6, pp. 921–928, 2009.
30. L. Taiz and E. Zeiger, Plant Physiology, Sinauer Associates, Sunderland, Mass, USA, 2002.
31. A. Schaller and T. Diez, "Plant specific aspects of heavy metal uptake and comparison with quality standards for food and forage crops," in Der Einfluß von festen Abfällen auf Böden, Pflanzen, D. Sauerbeck and S. Lübben, Eds., pp. 92–125, KFA, Jülich, Germany, 1991, (German).

32. M. G. Kibra, "Effects of mercury on some growth parameters of rice (Oryza sativaL.)," Soil & Environment, vol. 27, no. 1, pp. 23–28, 2008.
33. I. Ahmad, M. J. Akhtar, Z. A. Zahir, and A. Jamil, "Effect of cadmium on seed germination and seedling growth of four wheat (Triticum aestivum L.) cultivars,"Pakistan Journal of Botany, vol. 44, no. 5, pp. 1569–1574, 2012.
34. A. Kabata-Pendias, Trace Elements in Soils and Plants, CRC Press, Boca Raton, Fla, USA, 3rd edition, 2001.
35. A. R. Marin, S. R. Pezeshki, P. H. Masscheleyn, and H. S. Choi, "Effect of dimethylarsinic acid (DMAA) on growth, tissue arsenic and photosynthesis of rice plants," Journal of Plant Nutrition, vol. 16, no. 5, pp. 865–880, 1993.
36. M. J. Abedin, J. Cotter-Howells, and A. A. Meharg, "Arsenic uptake and accumulation in rice (Oryza sativa L.) irrigated with contaminated water," Plant and Soil, vol. 240, no. 2, pp. 311–319, 2002.
37. A. C. Barrachina, F. B. Carbonell, and J. M. Beneyto, "Arsenic uptake, distribution, and accumulation in tomato plants: effect of arsenite on plant growth and yield,"Journal of Plant Nutrition, vol. 18, no. 6, pp. 1237–1250, 1995.
38. M. S. Cox, P. F. Bell, and J. L. Kovar, "Differential tolerance of canola to arsenic when grown hydroponically or in soil," Journal of Plant Nutrition, vol. 19, no. 12, pp. 1599–1610, 1996.
39. M. S. Yourtchi and H. R. Bayat, "Effect of cadmium toxicity on growth, cadmium accumulation and macronutrient content of durum wheat (Dena CV.)," International Journal of Agriculture and Crop Sciences, vol. 6, no. 15, pp. 1099–1103, 2013.
40. W. Jiang, D. Liu, and W. Hou, "Hyperaccumulation of cadm ium by roots, bulbs and shoots of garlic," Bioresource Technology, vol. 76, no. 1, pp. 9–13, 2001.
41. M. Wang, J. Zou, X. Duan, W. Jiang, and D. Liu, "Cadmium accumulation and its effects on metal uptake in maize (Zea mays L.)," Bioresource Technology, vol. 98, no. 1, pp. 82–88, 2007.
42. K. Jayakumar, M. Rajesh, L. Baskaran, and P. Vijayarengan, "Changes in nutritional metabolism of tomato (Lycopersicon esculantum Mill.) plants exposed to increasing concentration

of cobalt chloride," International Journal of Food Nutrition and Safety, vol. 4, no. 2, pp. 62–69, 2013.
43. K. Jayakumar, C. A. Jaleel, and M. M. Azooz, "Phytochemical changes in green gram (Vigna radiata) under cobalt stress," Global Journal of Molecular Sciences, vol. 3, no. 2, pp. 46–49, 2008.
44. K. Jayakumar, C. A. Jaleel, and P. Vijayarengan, "Changes in growth, biochemical constituents, and antioxidant potentials in radish (Raphanus sativus L.) under cobalt stress," Turkish Journal of Biology, vol. 31, no. 3, pp. 127–136, 2007.
45. D. C. Sharma and C. P. Sharma, "Chromium uptake and its effects on growth and biological yield of wheat," Cereal Research Communications, vol. 21, no. 4, pp. 317–322, 1993.
46. S. K. Panda and H. K. Patra, "Nitrate and ammonium ions effect on the chromium toxicity in developing wheat seedlings," Proceedings of the National Academy of Sciences, India, vol. 70, pp. 75–80, 2000.
47. R. Moral, J. Navarro Pedreno, I. Gomez, and J. Mataix, "Effects of chromium on the nutrient element content and morphology of tomato," Journal of Plant Nutrition, vol. 18, no. 4, pp. 815–822, 1995.
48. R. Moral, I. Gomez, J. N. Pedreno, and J. Mataix, "Absorption of Cr and effects on micronutrient content in tomato plant (Lycopersicum esculentum M.)," Agrochimica, vol. 40, no. 2-3, pp. 132–138, 1996.
49. N. Nematshahi, M. Lahouti, and A. Ganjeali, "Accumulation of chromium and its effect on growth of (Allium cepa cv. Hybrid)," European Journal of Experimental Biology, vol. 2, no. 4, pp. 969–974, 2012.
50. C. M. Cook, A. Kostidou, E. Vardaka, and T. Lanaras, "Effects of copper on the growth, photosynthesis and nutrient concentrations of Phaseolus plants,"Photosynthetica, vol. 34, no. 2, pp. 179–193, 1997.
51. C. Kjær and N. Elmegaard, "Effects of copper sulfate on black bindweed (Polygonum convolvulus L.)," Ecotoxicology and Environmental Safety, vol. 33, no. 2, pp. 110–117, 1996.
52. A. R. Sheldon and N. W. Menzies, "The effect of copper toxicity on the growth and root morphology of Rhodes grass (Chloris

gayana Knuth.) in resin buffered solution culture," Plant and Soil, vol. 278, no. 1-2, pp. 341–349, 2005.

53. X. Du, Y.-G. Zhu, W.-J. Liu, and X.-S. Zhao, "Uptake of mercury (Hg) by seedlings of rice (Oryza sativa L.) grown in solution culture and interactions with arsenate uptake," Environmental and Experimental Botany, vol. 54, no. 1, pp. 1–7, 2005.
54. C. H. C. Shekar, D. Sammaiah, T. Shasthree, and K. J. Reddy, "Effect of mercury on tomato growth and yield attributes," International Journal of Pharma and Bio Sciences, vol. 2, no. 2, pp. B358–B364, 2011.
55. S. K. Arya and B. K. Roy, "Manganese induced changes in growth, chlorophyll content and antioxidants activity in seedlings of broad bean (Vicia faba L.)," Journal of Environmental Biology, vol. 32, no. 6, pp. 707–711, 2011.
56. Z. Asrar, R. A. Khavari-Nejad, and H. Heidari, "Excess manganese effects on pigments of Mentha spicata at flowering stage," Archives of Agronomy and Soil Science, vol. 51, no. 1, pp. 101–107, 2005.
57. S. Doncheva, K. Georgieva, V. Vassileva, Z. Stoyanova, N. Popov, and G. Ignatov, "Effects of succinate on manganese toxicity in pea plants," Journal of Plant Nutrition, vol. 28, no. 1, pp. 47–62, 2005.
58. M. Shenker, O. E. Plessner, and E. Tel-Or, "Manganese nutrition effects on tomato growth, chlorophyll concentration, and superoxide dismutase activity," Journal of Plant Physiology, vol. 161, no. 2, pp. 197–202, 2004.
59. I. S. Sheoran, H. R. Singal, and R. Singh, "Effect of cadmium and nickel on photosynthesis and the enzymes of the photosynthetic carbon reduction cycle in pigeonpea (Cajanus cajan L.)," Photosynthesis Research, vol. 23, no. 3, pp. 345–351, 1990.
60. B. Y. Khalid and J. Tinsley, "Some effects of nickel toxicity on rye grass," Plant and Soil, vol. 55, no. 1, pp. 139–144, 1980.
61. T. Pandolfini, R. Gabbrielli, and C. Comparini, "Nickel toxicity and peroxidase activity in seedlings of Triticum aestivum L.," Plant, Cell and Environment, vol. 15, no. 6, pp. 719–725, 1992.
62. V. S. Barsukova and O. I. Gamzikova, "Effects of nickel surplus on the element content in wheat varieties contrasting in Ni resistance," Agrokhimiya, vol. 1, pp. 80–85, 1999.

63. Y.-C. Lin and C.-H. Kao, "Nickel toxicity of rice seedlings: Cell wall peroxidase, lignin, and $NiSO_4$-inhibited root growth," Crop, Environment Bioinformatics, vol. 2, pp. 131–136, 2005.
64. A. Hussain, N. Abbas, F. Arshad et al., "Effects of diverse doses of lead (Pb) on different growth attributes of Zea mays L.," Agricultural Sciences, vol. 4, no. 5, pp. 262–265, 2013.
65. M. Kabir, M. Z. Iqbal, and M. Shafiq, "Effects of lead on seedling growth of Thespesia populnea L.," Advances in Environmental Biology, vol. 3, no. 2, pp. 184–190, 2009.
66. M. Moustakas, T. Lanaras, L. Symeonidis, and S. Karataglis, "Growth and some photosynthetic characteristics of field grown Avena sativa under copper and lead stress," Photosynthetica, vol. 30, no. 3, pp. 389–396, 1994.
67. R. Manivasagaperumal, S. Balamurugan, G. Thiyagarajan, and J. Sekar, "Effect of zinc on germination, seedling growth and biochemical content of cluster bean (Cyamopsis tetragonoloba (L.) Taub)," Current Botany, vol. 2, no. 5, pp. 11–15, 2011.
68. S. Doncheva, Z. Stoynova, andV. Velikova, "Influence of succinate on zinc toxicity of pea plants," Journal of Plant Nutrition, vol. 24, no. 6, pp. 789–804, 2001.
69. M. Bonnet, O. Camares, and P. Veisseire, "Effects of zinc and influence ofAcremonium lolii on growth parameters, chlorophyll a fluorescence and antioxidant enzyme activities of ryegrass (Lolium perenne L. cv Apollo)," Journal of Experimental Botany, vol. 51, no. 346, pp. 945–953, 2000.
70. A. M. Nicholls and T. K. Mal, "Effects of lead and copper exposure on growth of an invasive weed, Lythrum salicaria L. (Purple Loosestrife)," Ohio Journal of Science, vol. 103, no. 5, pp. 129–133, 2003.
71. A. Ghani, "Toxic effects of heavy metals on plant growth and metal accumulation in maize (Zea mays L.)," Iranian Journal of Toxicology, vol. 3, no. 3, pp. 325–334, 2010.
72. A. J. M. Baker, "Accumulators and excluders strategies in the response of plants to heavy metals," Journal of Plant Nutrition, vol. 3, pp. 643–654, 1981.
73. M. J. Blaylock, D. E. Salt, S. Dushenkov et al., "Enhanced accumulation of Pb in Indian mustard by soil-applied chelating

agents," Environmental Science and Technology, vol. 31, no. 3, pp. 860–865, 1997.

74. M. E. V. Schmoger, M. Oven, and E. Grill, "Detoxification of arsenic by phytochelatins in plants," Plant Physiology, vol. 122, no. 3, pp. 793–801, 2000.
75. C. Garbisu and I. Alkorta, "Bioremediation: principles and future," Journal of Clean Technology, Environmental Toxicology and Occupational Medicine, vol. 6, no. 4, pp. 351–366, 1997.
76. C. Garbisu and I. Alkorta, "Basic concepts on heavy metal soil bioremediation," The European Journal of Mineral Processing and Environmental Protection, vol. 3, no. 1, pp. 58–66, 2003.
77. P. Wang, T. Mori, K. Komori, M. Sasatsu, K. Toda, and H. Ohtake, "Isolation and characterization of an Enterobacter cloacae strain that reduces hexavalent chromium under anaerobic conditions," Applied and Environmental Microbiology, vol. 55, no. 7, pp. 1665–1669, 1989.
78. Y. Ishibashi, C. Cervantes, and S. Silver, "Chromium reduction in Pseudomonas putida," Applied and Environmental Microbiology, vol. 56, no. 7, pp. 2268–2270, 1990.
79. C. Garbisu, M. J. Llama, and J. L. Serra, "Effect of heavy metals on chromate reduction by Bacillus subtilis," Journal of General and Applied Microbiology, vol. 43, no. 6, pp. 369–371, 1997.
80. C. Garbisu, I. Alkorta, M. J. Llama, and J. L. Serra, "Aerobic chromate reduction byBacillus subtilis," Biodegradation, vol. 9, no. 2, pp. 133–141, 1998.
81. C. Garbisu, S. González, W.-H. Yang et al., "Physiological mechanisms regulating the conversion of selenite to elemental selenium by Bacillus subtilis," BioFactors, vol. 5, no. 1, pp. 29–37, 1995.
82. R. Ajaz Haja Mohideena, V. Thirumalai Arasuc, K. R. Narayananb, and M. I. Zahir Hussaind, "Bioremediation of heavy metal contaminated soil by the exigobacterium and accumulation of Cd, Ni, Zn and Cu from soil environment," International Journal of Biological Technology, vol. 1, no. 2, pp. 94–101, 2010.
83. D. van der Lelie, P. Corbisier, L. Diels et al., "The role of bacte ria in the phytoremediation of heavy metals," in Phytoremediation of Contaminated Soil and Water, N. Terry and E. Banuelos, Eds., pp. 265–281, G Lewis, Boca Raton, Fla, USA, 1999.

84. M. Huyer and W. J. Page, "Zn^{2+} increases siderophore production in Azotobacter vinelandii," Applied and Environmental Microbiology, vol. 54, no. 11, pp. 2625–2631, 1988.
85. C. White, A. K. Sharman, and G. M. Gadd, "An integrated microbial process for the bioremediation of soil contaminated with toxic metals," Nature Biotechnology, vol. 16, no. 6, pp. 572–575, 1998.
86. J. L. Hobman and N. L. Brown, "bacterial mercury-resistance genes," Metal ions in biological systems, vol. 34, pp. 527–568, 1997.
87. D. R. Lovley and J. R. Lloyd, "Microbes with a mettle for bioremediation," Nature Biotechnology, vol. 18, no. 6, pp. 600–601, 2000.
88. M. Valls, S. Atrian, V. de Lorenzo, and L. A. Fernández, "Engineering a mouse metallothionein on the cell surface of Ralstonia eutropha CH34 for immobilization of heavy metals in soil," Nature Biotechnology, vol. 18, no. 6, pp. 661–665, 2000.
89. M. Urgun-Demirtas, B. Stark, and K. Pagilla, "Use of genetically engineered microorganisms (GEMs) for the bioremediation of contaminants," Critical Reviews in Biotechnology, vol. 26, no. 3, pp. 145–164, 2006.
90. O. P. Abioye, "Biological remediation of hydrocarbon and heavy metals contaminated soil," in Soil Contamination, S. Pascucci, Ed., InTech, Vienna, Austria, 2011.
91. A. McCauley, C. Jones, and J. Jacobsen, "Soil pH and organic matter," in Nutrient Management Module, vol. 8, Montana State University Extension, Bozeman, Mont, USA, 2009.
92. A. Karaca, "Effect of organic wastes on the extractability of cadmium, copper, nickel, and zinc in soil," Geoderma, vol. 122, no. 2–4, pp. 297–303, 2004.
93. T. Namgay, B. Singh, and B. P. Singh, "Influence of biochar application to soil on the availability of As, Cd, Cu, Pb, and Zn to maize (Zea mays L.)," Soil Research, vol. 48, no. 6-7, pp. 638–647, 2010.
94. J. M. Novak, W. J. Busscher, D. L. Laird, M. Ahmedna, D. W. Watts, and M. A. S. Niandou, "Impact of biochar amendment on fertility of a southeastern coastal plain soil," Soil Science, vol. 174, no. 2, pp. 105–112, 2009.

95. D. E. Salt, R. D. Smith, and I. Raskin, "Phytoremediation," Annual Review of Plant Biology, vol. 49, pp. 643–668, 1998.
96. R. L. Chaney, M. Malik, Y. M. Li et al., "Phytoremediation of soil metals," Current Opinion in Biotechnology, vol. 8, no. 3, pp. 279–284, 1997.
97. A. J. M. Baker and R. R. Brooks, "Terrestrial higher plants which hyperaccumulate metallic elements: a review of their distribution, ecology and phytochemistry,"Biorecovery, vol. 1, pp. 81–126, 1989.
98. S. P. McGrath and F. Zhao, "Phytoextraction of metals and metalloids from contaminated soils," Current Opinion in Biotechnology, vol. 14, no. 3, pp. 277–282, 2003.
99. R. D. Reeves and A. J. M. Baker, "Metal-accumulating plants," in Phytoremediation of Toxic Metals: Using Plants to Clean Up the Environment, I. Raskin and B. D. Ensley, Eds., pp. 193–229, Wiley, New York, NY, USA, 2000.
100. L. Q. Ma, K. M. Komar, C. Tu, W. Zhang, Y. Cai, and E. D. Kenelley, "A fern that hyperaccumulates arsenic—a hardy, versatile, fast-growing plant helps to remove arsenic from contaminated soils," Nature, vol. 409, p. 579, 2001.
101. X. E. Yang, X. X. Long, H. B. Ye, Z. L. He, D. V. Calvert, and P. J. Stoffella, "Cadmium tolerance and hyperaccumulation in a new Zn-hyperaccumulating plant species (Sedum alfredii Hance)," Plant and Soil, vol. 259, no. 1-2, pp. 181–189, 2004.
102. F. Navari-Izzo and M. F. Quartacci, "Phytoremediation of metals," Minerva Biotecnologica, vol. 13, no. 2, pp. 73–83, 2001.
103. L. Van Ginneken, E. Meers, R. Guisson et al., "Phytoremediation for heavy metal-contaminated soils combined with bioenergy production," Journal of Environmental Engineering and Landscape Management, vol. 15, no. 4, pp. 227–236, 2007.
104. S. D. Ebbs and L. V. Kochian, "Toxicity of zinc and copper to Brassica species: implications for phytoremediation," Journal of Environmental Quality, vol. 26, no. 3, pp. 776–781, 1997.
105. R. L. Chaney, S. L. Brown, L. Yin-Ming et al., "Progress in risk assessment for soil metals, and in-situ remediation and phytoextraction of metals from hazardous contaminated soils," in Proceedings of the US EPA's Conference Phytoremediation: State of the Science Conference, Boston, Mass, USA, 2000.

106. Y. Chen, X. Li, and Z. Shen, "Leaching and uptake of heavy metals by ten different species of plants during an EDTA-assisted phytoextraction process," Chemosphere, vol. 57, no. 3, pp. 187–196, 2004.
107. H. Lai and Z. Chen, "The EDTA effect on phytoextraction of single and combined metals-contaminated soils using rainbow pink (Dianthus chinensis)," Chemosphere, vol. 60, no. 8, pp. 1062–1071, 2005.
108. S. C. Wu, K. C. Cheung, Y. M. Luo, and M. H. Wong, "Effects of inoculation of plant growth-promoting rhizobacteria on metal uptake by Brassica juncea," Environmental Pollution, vol. 140, no. 1, pp. 124–135, 2006.
109. K. K. Chiu, Z. H. Ye, and M. H. Wong, "Growth of Vetiveria zizanioides andPhragmities australis on Pb/Zn and Cu mine tailings amended with manure compost and sewage sludge: a greenhouse study," Bioresource Technology, vol. 97, no. 1, pp. 158–170, 2006.
110. E. Lombi, F. J. Zhao, S. J. Dunham, and S. P. McGrath, "Phytoremediation of heavy metal-contaminated soils: Natural hyperaccumulation versus chemically enhanced phytoextraction," Journal of Environmental Quality, vol. 30, no. 6, pp. 1919–1926, 2001.
111. C. D. Jadia and M. H. Fulekar, "Phytotoxicity and remediation of heavy metals by fibrous root grass (sorghum)," Journal of Applied Biosciences, vol. 10, pp. 491–499, 2008.
112. V. Laperche, S. J. Traina, P. Gaddam, and T. J. Logan, "Effect of apatite amendments on plant uptake of lead from contaminated sail," Environmental Science and Technology, vol. 30, no. 10, pp. 1540–1552, 1997.
113. A. P. G. C. Marques, R. S. Oliveira, A. O. S. S. Rangel, and P. M. L. Castro, "Application of manure and compost to contaminated soils and its effect on zinc accumulation by Solanum nigrum inoculated with arbuscular mycorrhizal fungi,"Environmental Pollution, vol. 151, no. 3, pp. 608–620, 2008.
114. D. C. Adriano, W. W. Wenzel, J. Vangronsveld, and N. S. Bolan, "Role of assisted natural remediation in environmental cleanup," Geoderma, vol. 122, no. 2–4, pp. 121–142, 2004.

115. United States Environmental Protection Agency, Electrokinetic and Phytoremediation In Situ Treatment of Metal-Contaminated Soil: State-of-the-Practice, EPA/542/R-00/XXX, Environmental Protection Agency, Office of Solid Waste and Emergency Response Technology Innovation Office, Washington, DC, USA, 2000.
116. I. Raskin and B. D. Ensley, Phytoremediation of Toxic Metals: Using Plants to Clean Up the Environment, John Wiley & Sons, New York, NY, USA, 2000.
117. C. L. Rugh, J. F. Senecoff, R. B. Meagher, and S. A. Merkle, "Development of transgenic yellow poplar for mercury phytoremediation," Nature Biotechnology, vol. 16, no. 10, pp. 925–928, 1998.
118. R. B. Meagher, C. L. Rugh, M. K. Kandasamy, G. Gragson, and N. J. Wang, "Engineered phytoremediation of mercury pollution in soil and water using bacterial genes," in Phytoremediation of Contaminated Soil and Water, N. Terry and G. Bañuelos, Eds., pp. 201–219, Lewis Publishers, Boca Raton, Fla, USA, 2000.
119. United States Environmental Protection Agency (USEPA), "Introduction to phytoremediation," EPA 600/R-99/107, U.S. Environmental Protection Agency, Office of Research and Development, Cincinnati, Ohio, USA, 2000.
120. R. B. Meagher, "Phytoremediation: An Affordable, Friendly Technology to Restore Marginal Lands in the Twenty-First Century," 1998, http://www.lsc.psu.edu/nas/Panelists/Meagher%20comment.html.
121. N. Terry, A. M. Zayed, M. P. de Souza, and A. S. Tarun, "Selenium in higher plants,"Annual Review of Plant Biology, vol. 51, pp. 401–432, 2000.
122. G. S. Bañuelos, H. A. Ajwa, B. Mackey et al., "Evaluation of different plant species used for phytoremediation of high soil selenium," Journal of Environmental Quality, vol. 26, no. 3, pp. 639–646, 1997.
123. N. Weyens, D. van der Lelie, S. Taghavi, L. Newman, and J. Vangronsveld, "Exploiting plant-microbe partnerships to improve biomass production and remediation," Trends in Biotechnology, vol. 27, no. 10, pp. 591–598, 2009.

124. E. J. Joner and C. Leyval, "Time-course of heavy metal uptake in maize and clover as affected by root density and different mycorrhizal inoculation regimes," Biology and Fertility of Soils, vol. 33, no. 5, pp. 351–357, 2001.
125. A. Jamal, N. Ayub, M. Usman, and A. G. Khan, "Arbuscular mycorrhizal fungi enhance zinc and nickel uptake from contaminated soil by soybean and lentil,"International Journal of Phytoremediation, vol. 4, no. 3, pp. 205–221, 2002.
126. A. P. G. C. Marques, R. S. Oliveira, A. O. S. S. Rangel, and P. M. L. Castro, "Zinc accumulation in Solanum nigrum is enhanced by different arbuscular mycorrhizal fungi," Chemosphere, vol. 65, no. 7, pp. 1256–1263, 2006.
127. A. Heggo, J. S. Angle, and R. L. Chaney, "Effects of vesicular-arbuscular mycorrhizal fungi on heavy metal uptake by soybeans," Soil Biology & Biochemistry, vol. 22, no. 6, pp. 865–869, 1990.
128. M. Janoušková, D. Pavlíková, and M. Vosátka, "Potential contribution of arbuscular mycorrhiza to cadmium immobilisation in soil," Chemosphere, vol. 65, no. 11, pp. 1959–1965, 2006.
129. L. A. Harrier and C. A. Watson, "The potential role of arbuscular mycorrhizal (AM) fungi in the bioprotection of plants against soil-borne pathogens in organic and/or other sustainable farming systems," Pest Management Science, vol. 60, no. 2, pp. 149–157, 2004.
130. I. M. Cardoso and T. W. Kuyper, "Mycorrhizas and tropical soil fertility," Agriculture, Ecosystems and Environment, vol. 116, no. 1-2, pp. 72–84, 2006.
131. S. F. Wright, V. S. Green, and M. A. Cavigelli, "Glomalin in aggregate size classes from three different farming systems," Soil & Tillage Research, vol. 94, no. 2, pp. 546–549, 2007.
132. G. U. Chibuike, "Use of mycorrhiza in soil remediation: a review," Scientific Research and Essays, vol. 8, no. 35, pp. 1679–1687, 2013.
133. G. Díaz, C. Azcón-Aguilar, and M. Honrubia, "Influence of arbuscular mycorrhizae on heavy metal (Zn and Pb) uptake and growth of Lygeum spartum and Anthyllis cytisoides," Plant and Soil, vol. 180, no. 2, pp. 241–249, 1996.

134. E. J. Joner and C. Leyval, "Uptake of ^{109}Cd by roots and hyphae of a Glomus mosseae/Trifolium subterraneum mycorrhiza from soil amended with high and low concentrations of cadmium," New Phytologist, vol. 135, no. 2, pp. 353–360, 1997.

135. C. C. Chao and Y. P. Wang, "Effects of heavy-metals on the infection of vesicular arbuscular mycorrhizae and the growth of maize," Journal of the Agricultural Association of China, vol. 152, pp. 34–45, 1990.

136. C. Del Val, J. M. Barea, and C. Azcón-Aguilar, "Diversity of arbuscular mycorrhizal fungus populations in heavy-metal-contaminated soils," Applied and Environmental Microbiology, vol. 65, no. 2, pp. 718–723, 1999.

137. I. Weissenhorn and C. Leyval, "Spore germination of arbuscular mycorrhizal fungi in soils differing in heavy metal content and other parameters," European Journal of Soil Biology, vol. 32, no. 4, pp. 165–172, 1996.

138. B. R. Glick, D. M. Karaturovic, and P. C. Newell, "A novel procedure for rapid isolation of plant growth promoting pseudomonads," Canadian Journal of Microbiology, vol. 41, no. 6, pp. 533–536, 1995.

139. A. A. Kamnev and D. van der Lelie, "Chemical and biological parameters as tools to evaluate and improve heavy metal phytoremediation," Bioscience Reports, vol. 20, no. 4, pp. 239–258, 2000.

140. A. G. Khan, "Role of soil microbes in the rhizospheres of plants growing on trace metal contaminated soils in phytoremediation," Journal of Trace Elements in Medicine and Biology, vol. 18, no. 4, pp. 355–364, 2005.

141. B. R. Glick, D. M. Penrose, and J. Li, "A model for the lowering of plant ethylene concentrations by plant growth-promoting bacteria," Journal of Theoretical Biology, vol. 190, no. 1, pp. 63–68, 1998.

142. M. L. E. Reed and B. R. Glick, "Growth of canola (Brassica napus) in the presence of plant growth-promoting bacteria and either copper or polycyclic aromatic hydrocarbons," Canadian Journal of Microbiology, vol. 51, no. 12, pp. 1061–1069, 2005.

143. X. Sheng and J. Xia, "Improvement of rape (Brassica napus) plant growth and cadmium uptake by cadmium-resistant bacteria," Chemosphere, vol. 64, no. 6, pp. 1036–1042, 2006.
144. S. Zaidi, S. Usmani, B. R. Singh, and J. Musarrat, "Significance of Bacillus subtilisstrain SJ-101 as a bioinoculant for concurrent plant growth promotion and nickel accumulation in Brassica juncea," Chemosphere, vol. 64, no. 6, pp. 991–997, 2006.
145. M. Madhaiyan, S. Poonguzhali, and S. A. Torgmin, "Metal tolerating methylotrophic bacteria reduces nickel and cadmium toxicity and promotes plant growth of tomato (Lycopersicon esculentum L.)," Chemosphere, vol. 69, no. 2, pp. 220–228, 2007.
146. A. Vivas, B. Biró, J. M. Ruíz-Lozano, J. M. Barea, and R. Azcón, "Two bacterial strains isolated from a Zn-polluted soil enhance plant growth and mycorrhizal efficiency under Zn-toxicity," Chemosphere, vol. 62, no. 9, pp. 1523–1533, 2006.

chapter 2

Coping Mechanisms of Plants to Metal Contaminated Soil

Melanie Mehes-Smith[1], Kabwe Nkongolo[2], and Ewa Cholewa[1]

[1]Department of Biology, Nipissing University, North Bay, Ontario, Canada

[2]Department of Biology, Laurentian University, Sudbury, Ontario, Canada

INTRODUCTION

Metals such as cobalt (Co), copper (Cu), iron (Fe) and nickel (Ni) are essential for normal plant growth and development since they contribute to the function of many enzymes and proteins. However, metals can potentially become toxic to plants when they are present at high levels in their bioavailable forms (Hall, 2002). Phytotoxic levels of one or more inorganic ions in soil can be found in various parts of the world. These toxic sites occurred through natural processes or by anthropogenic effects. Naturally toxic soils include saline, acidic and serpentine soils, while anthropogenic polluted soils occur through mining activities, aerial fallout, and the run-off from galvanized sources of electricity pylons or motorway verges polluted by vehicle exhaust fumes (Bradshaw, 1984). The biochemical effect of metals on plants varies and the excess metal usually results in oxidative damage which affects their phenotype (Kachout *et al.,* 2009)

Plants colonizing metal-contaminated soils are classified as resistant and have adapted to this stressed environment. Heavy metal resistance can be achieved by avoidance and/or tolerance. Avoiders are plants that are able to protect themselves by preventing metal ions from entering their cellular cytoplasm, while tolerant plants are able to detoxify metal ions that have crossed the plasma membrane or internal organelle biomembranes (Millaleo *et al.,* 2010). Based on strategies used by plants growing on metal-contaminated soils, Baker and Walker (1990) classified them into three categories; metal excluder, indicators and accumulators/hyperaccumulators. The excluder group includes the majority of plant species that limit the translocation of heavy metals and maintain low levels of contaminants in their aerial tissues over an extensive range of soil concentrations. Plants that are metal indicators accumulate metals in their harvestable biomass and these levels generally are reflective of the metal concentration in the soil. Metal accumulators/hyperaccumulators are plants that increase internal sequestration, translocation and accumulation of metals in their harvestable biomass to levels that far exceed those found in the soil (Mganga *et al.,* 2011; Baker and Walker 1990). Plants can accumulate and cope with the effects of high internal metal concentrations by the upregulation of the antioxidant defense system. This system is activated in order to respond to the deleterious effects caused by reactive oxygen species (Solanki and Dhankhar 2011).

Coping strategies allow the establishment of plant communities on metal contaminated soils. This is possible since some plants have adapted to these hostile sites by evolving mechanisms to deal with the toxic effects of metals in soil on plants. There is a need of identifying plants that are able to deal with excess metal in soil. Without these plants, the lands would remain barren and unsustainable.

The importance of plants in the remediation of heavy metal polluted soil is discussed in details in the present chapter. A review of the current knowledge on metal resistance mechanisms, as well as the potential genes and their role in metal homeostasis in plants will be examined. Finally, the coping mechanisms used by plants growing under metal contamination will be discussed.

REMEDIATION OF HEAVY METAL CONTAMINATED SITES

Soils that are heavily contaminated by metals may pose health risks to humans and to other living organisms in an ecosystem. Current techniques used to remediate metal contaminated soils include excavation, chemical stabilization, soil washing or soil flushing, but these methods are costly and impractical. There is a need to develop effective, low-cost and sustainable methods for soil bioremediation. The revegetation of these sites appears to be the most suitable method for long term land reclamation since plants can improve nutrient soil conditions. This can lead to the establishment of a self-sustaining vegetative cover, which in turn can prevent soil erosion (Wei *et al.*, 2005). Phytoremediation is an inexpensive and solar-driven approach that is performed *in situ*. It can be used to remove, stabilize and detoxify organic and inorganic pollutants including heavy metals from air, soil and liquid substrates (Salt *et al.*, 1998). An example of a reclaimed metal contaminated site in the mining region of Northern Ontario (Canada) is illustrated in figure 1. Plant species selected for land reclamation should grow and spread fast and be able to establish an effective soil cover. It is therefore important to search for plants that have spontaneously colonized these disturbed sites. Moreover, heavy metal contaminated mining sites exhibit physiochemical characteristics that are not suitable for the vast majority of plant species; hence the colonization of these sites is slow. However, plants that are resistant to this toxic environment can easily spread since there is a lack of competitors. It has been demonstrated that annual species have an extensive adaptive capacity compared to perennial genotypes due to their long-term natural selection (Wei *et al.*, 2005).

Phytoremediation is composed of five main subgroups: phytoextraction, phytovolatilisation phytostabilization, phytodegradation and rhizofiltration. Phytoextraction is a process by which plants extract metals from soil by accumulating them in their aerial biomass. These plants can be harvested and metals can be extracted from their tissues. Plants that accumulate metals in their aerial tissues have been involved in the phytoextraction of several metals including Cd, Cr, Cu, Hg, Pb, Ni Se and Zn (Yong and Ma 2002).

The accumulation of metals by plants is interesting from an environmental or agronomic point of view. In mining or industrial sites, as well as their surrounding areas, heavy metals are responsible for severe soil contamination. In these cases, accumulator plants could be used for phytoremediation as they are likely able to remove metals from soils (Salt *et al.,* 1998; Salt *et al.,* 1995). Since some heavy metals are also essential minerals that can be deficient in staple food crops, genetic determinants of hyperaccumulation could be utilized in biofortification to improve the nutritional value of these crops (Frérot *et al.,* 2010; Cakmak, 2008; Jeong and Guerinot 2008; Mayer *et al.,*2008).

Figure 1: A metal contaminated site in Sudbury, Ontario, Canada; a) before remediation and b) after remediation (photos courtesy of Keith Winterhalder and David Pearson from Laurentian University).

These metal accumulator plants could also convert metals and release them in a volatile form. This process is known as phytovolatilization. Phytostabilization on the other hand, is a method that uses plants with a low ability for metal uptake to stabilize the contaminated soil thus preventing erosion. This limits the metals from entering the food chain. Plants can also be utilized for phytodegradation since they can in combination with microorganisms degrade organic pollutants. Finally, rhizofiltration is a process by which plant roots absorb metals from waste streams (Pulford and Watson 2003;Dushenkov *et al.*, 1995).

RESISTANCE

Levitt (1980) stated that heavy metal contaminated environments act as stress factors on plants, which causes physiological reaction change that reduces or inhibits plant vigor and growth. A plant showing injury or death due to metal stress is deemed sensitive to its environment. On the other hand, resistant plants can survive and reproduce under metal stress conditions (Ernst *et al.*, 2008). In general, plants can achieve resistance to heavy metals by avoidance or tolerance.

Avoidance

Avoidance occurs when plants restrict the uptake of metals within root tissue by several strategies. In environments where the soil metal contamination is heterogeneously distributed, plants can prevent metal uptake by exploring less contaminated soil. Another avoidance strategy involves mycorrhizal fungi, where they can extend their hyphae outside the plants rooting zone up to several tens of meters and transfer the necessary elements to the plant (Ernst, 2006; Baker, 1987). Also, these metal tolerant fungi can increase plant metal resistance by changing the metals speciation or by restricting the metal transfer into the plant (Ernst, 2006). Arines *et al.* (1989) found that mycorrhizal *Trifolium pratense* (red clover) plants growing in acid soils had lower levels of Mn in their roots and shoots as compared to the non-mycorrhizal plants. Plants can also restrict contaminant uptake in root tissues by immobilizing metals for example through root exudates in the rhizosphere. A role of root exudates is to chelate metals and stop their entry inside the cell. The cell wall has also been found to be involved in restricting metal uptake into the cell's cytoplasm (Mganga *et al.*, 2011).

Tolerance

In the absence of avoidance strategies, some plants can grow and survive in soil contaminated with toxic levels of heavy metals which are otherwise lethal or detrimental for growth and survival of others genotypes of the same or of different species (Maestri and Marmiroli 2012). Plants exhibiting tolerance are internally protected from the stress of metals that have entered the cell's cytoplasm (Baker, 1987). Metallophytes (metal tolerant plant) can function normally even in the presence of higher plant-internal metal levels. Plants adapt to their environments by developing heritable tolerance mechanisms. Tolerance to specific metals has evolved independently several times in different species from local non tolerant ancestral plant populations (Schat *et al.,* 2000). Plants can exhibit tolerance to metals that are present in surplus in the soil. Each metal is under control of specific genes.

According to Bradshaw (1991) most species are in a state of genostasis. It is the restriction of genetic variability which limits the evolution of the population/species. In the absence of avoidance pathways, metal contaminated soil acts as a selection force on a population, where only the plants with tolerant genotype can survive and reproduce. This leads to a bottleneck, where few individuals survive and reproduce. In turn, metal tolerant populations can evolve rapidly following a disturbance such as contamination of soil with heavy metals. Plant adaptation to these sites occurs in populations for which tolerance variability already exists prior to the contamination (Maestri and Marmiroli 2012;Baker, 1987). Genes for the tolerance of metals are pre-existing at a low frequency in non tolerant populations of certain plant species (Ernst, 2006; Macnair, 1987).

Variation in Tolerance and Accumulation Characteristics

Variation occurs between species, populations and clones for tolerance and accumulation of metals.Assunçãno *et al.* (2003) found differences in the degree of chlorosis and concentration of metal for*Thlaspi caerulescens* (currently named *Noccaea caerulescens*) populations when grown in hydroponic solutions containing various Ni, Cd and Zn concentrations. Visioli *et al.* (2010) found differences in growth, morphology and Ni accumulation capacity when the Ni hyperaccumulator *T. caerulescens* and the non-metal adapted *T. caerluescens* were exposed to different Ni concentrations in hydroponic solutions. Besnard *et al.* (2009) used cleaved amplified polymorphic sites (CAPs) and microsatellites to determine the genetic variation for *T.*

caerulescens populations from metalliferous and non-metalliferous sites from Switzerland. They found a correlation between the level of heavy metals in soil and the variation at the target loci for the genes involved in encoding metal transporters.Basic *et al.* (2006) found similar results when they analyzed the genetic variation of different *T. caerulescens* population sampled from different soil types with single nucleotide polymorphism (SNPs) in target and non target genes. These results were also observed in *Populus spp.* Marmiroli *et al.* (2011) compared *Populus* clones and found variation in their capacity to accumulate Cd. This variation in Cd accumulation between clones was correlated with SNPs at some target genes. These results imply that gene flow is limited between individuals found on metal contaminated and those from uncontaminated sites, at least for the loci that are involved in the fitness of the individuals (Visioli *et al.,* 2012). On metal-enriched soil, there is a strong selection of local offspring, which conserves the metal tolerant genotypes (Ernst, 2006).

Genetics of Tolerance to Metals

Identifying genes involved in a specific adaptation is challenging. Metal tolerance and accumulation in plants are complex genetic systems. Plants have to modify their physiological processes in order to be able to survive in the environment in which they have germinated. In turn, the survival of a population to the contaminated environment is dependent on the inheritance of favourable traits. Tolerance mechanisms are heritable and variable, resulting from genes and gene products (Maestri and Marmiroli 2012). Variation in the evolution of metal tolerance exists over species, populations and clones (Baker, 1987). Some species do not show variation in tolerance and accumulations. In order to determine genes involved in metal tolerance and accumulation, segregating analyses were used, where parents with contrasting phenotypes were crossed to produce progeny. Studies have determined that in many species, metal tolerance and accumulation are genetically independent (Assunçãno *et al.,* 2006). For example, in *Arabidopsis halleri*, Cd tolerance and accumulation segregated as independent traits while Cd and Zn tolerance and accumulation cosegretated. In this later species, two or more genes were proposed to be involved in Cd and Zn accumulation but only one gene for Cd and Zn tolerance (Bert *et al.,* 2003; Bert *et al.,* 2002).

In *T. caerulescens*, no genes involved in Cd, Zn and Ni tolerance and accumulation cosegregated. This suggests that there is a high probability that the genetic and physiological mechanisms for these traits are distinct from each other (Yang *et al.*, 2005a; Maestri *et al.*, 2010; Richau and Schat 2009; Assunçãno *et al.*, 2006; Zha *et al.*,2004). As a result, it is not possible to conclude that a plant with high levels of metals in aerial biomass is also metal tolerant. The concentration of metals in above ground tissue serves as an indication of the plant's potential metal tolerance (Frérot *et al.*, 2010).

Several techniques have been used to isolate and identify genes involved in heavy metal tolerance in plants, one of which is the quantitative trait loci (QTL) mapping. QTL mapping is a powerful tool in examining complex adaptive traits and in determining the number of genes involved in a trait as well as the genes effects and their interactions (Willems *et al.*, 2007). By mapping QTLs, it can be possible to identify or validate candidate genes involved in a complex trait such as metal tolerance and accumulation (Willems *et al.*, 2007). Other techniques used to identify genes for metal tolerance and accumulation are functional complementation in yeast mutants defective in metal homeostasis. These methods use plant cDNA expression libraries, as well as the identification of hypothesized pathways based on sequence similarities with plant cDNA libraries and genomic sequences (Lal, 2010). Transcriptome analyses have also been used to reveal genes involved in hyperaccumulation by analysing the differences in expression profiles or regulation-level of hyperaccumulator and non hyperaccumulator plants (Colzi *et al.*, 2011).

Few specific major genes have been found for Cd, Cu, Ni and Zn tolerance in *Silene vulgaris* by crossing plants from a metalliferous site with non tolerant plant from a nonmetalliferous site (Schat *et al.*, 1996; Schat *et al.*, 1993). Similar results were reported for Cu tolerance for *Mimulus guttatus*(Macnair, 1993) and Zn tolerance for *Arabidipsis halleri* (Bert *et al.*, 2003). In *S. vulgaris* and *M. guttatus*, modifier genes (minor genes) were involved in Cu tolerance, thus increasing tolerance and enhancing the effect of the major gene(s) (Smith and Macnair 1998). Only two QTL were involved in Ni accumulation and tolerance in *S. vulgaris* (Bratteler, 2005).

Studies aiming at identifying associations between molecular markers and metal tolerance and accumulation trait have been performed using interspecific and intraspecific crosses. When a high Zn accumulating *T. caerulescens* parent was crossed with a low Zn accumulating parent, two major QTLs were found to be involved in the increased of Zn accumulation in root (Assunçãno *et al.*,2006). Deniau *et al.*, (2006) performed QTL mapping for the hyperaccumulation of Zn and Cd in *T. caerulescens*. They found two QTLs responsible for Cd and two for Zn accumulation in roots. In addition, one QTL for Cd and three QTLs for Zn accumulation in shoot were characterized. Macnair*et al.* (1999) reported a major gene involved in Zn tolerance from the analysis of F_2 progeny derived from a cross between *A. halleri* (tolerant parent) and *Arabidopsis lyrata* (sensitive parent). Willems*et al.* (2007) generated a backcross progeny from the interspecific cross between *A. halleri* (tolerant parent) and *A. lyrata* (sensitive parent) and identified three major additive QTLs involved in Zn tolerance in *A. halleri*. These QTLs were mapped to three different chromosomes (3, 4 and 6) and colocalized with genes that have been known to be involved in metal tolerance and accumulation.*HMA4* (*Heavy Metal ATPase 4)* encodes a P-type ATPase pump localized at the plasmamembrane involved in loading Zn and Cd into the xylem. *MTP1*-A and *MTP1-B* are Metal Tolerance Protein- vacuolar transporters that are involved in Zn tolerance) (Gustin *et al.*, 2009; Krämer, 2005). Three new QTLs were identified and mapped to chromosomes 4, 6 and 7 by Filatov *et al.* 2007, when F_2progenies from a similar interspecific cross were analyzed. Frérot *et al.* (2010) also found Zn accumulation to be polygenic using *A. halleri* X *A. lyrata petraea* progenies. They determined that Zn accumulation is controlled by two QTLs in low Zn concentration and three QTLs in high Zn concentration. Four of the five QTLs mapped for Zn accumulation in their study were also reported in previous studies using *A. halleri* X *A. lyrata petraea* progenies (Frérot *et al.*, 2010; Filatov *et al.*,2007; Filatov *et al.*, 2006).

Courbot *et al.* (2007), also using progeny from interspecific cross between *A. halleri* and *A. lyarata*determined three QTLs involved in Cd tolerance. A major QTL region was found to be common to Cd (Courbot et al 2007) and Zn (Willems et al 2007) tolerance and was colocalized with the *HMA4*gene. Hanikenne et al (2008) identified the role of *HMA4* using RNAi-mediated silencing. They reported that when the expression of *HMA4* was down-regulated, less Zn was translocated from the root to the shoot. When this gene was expressed in *A. thaliana*, an increase in Zn translocation to aerial tissue was observed. This increase in Zn translocation in *A. thaliana* plants resulted in signs of Zn hypersensitivity. Therefore, the expression of *AhHMA4* alone was not adequate for Zn detoxification. Additional genes are involved in the *A. halleri* Zn hyperaccumulation (Hanikenne and Nouet 2011; Frérot *et al.,* 2010).

Using segregating progeny resulting from intraspecific crosses between a high Cd accumulating parent and a low Cd accumulating parent for *Glycine max*, Benitez *et al.* (2010) identified a major QTL in seeds that was named cd1. This gene was mapped on chromosome 9. Jegadeesan *et al.* (2010) also identified a major QTL, *cda1*, associated with Cd accumulation in seeds of *G. max*. These two major QTLs were mapped to the same region of chromosome 9 which suggested that cd1 and *cda1* may be identical. Both QTLs were found to be a dominant major gene involved in the control of low Cd uptake. By analyzing the *G. max* genome, Benitez *et al.* (2010) revealed that the cd1 QTL is localized in the vicinity of the P_{1B}-ATPase gene (designated as *GmHMA1)* and proposed that this gene is involved in the transport of Cd. Benitez *et al.* (2012) found a single-base substitution between two cultivars, Harosoy (high Cd content in seed Cd) and Fukuyutaka (low Cd content in seed) in this P_{1B}–ATPase gene. This mutation resulted in an amino acid substitution (glycine in Fukuyutaka and glutamic acid in Harosoy) in GmHMA1a. Since the glycine residue at the amino acid substitution site was conserved in AtHMA3, AtHMA4, AtHMA6 and AtHMA7, it was suggested that the GmHMA1a from Fukuyutaka was the wild type, responsible for low Cd accumulation in seed (Benitez *et al.,*2012). A dominant major gene involved in the control of Cd uptake was also observed in wheat (*Triticum aestivum*) (Clarke *et al.,* 1997) and oat (*Avena sativa* L.) (Tanhuanpää *et al.,* 2007). QTL analyses have also been performed in radish (*Raphanus sativus* L.). Xu *et al.* (2012) found a major QTL and three minor QTLs responsible for Cd accumulation in radish roots which were mapped on linkage groups 1, 4, 6 and 9. Induri *et al.* (2012) identified major QTLs for Cd response in *Populus*by performing a pseudo-backcross pedigree of *Populus trichocarpa* Torr. & Gray and *Populus deltoides* Bart. These QTLs were mapped to two different linkage groups. They performed a whole-genome microarray study and they were able to identify nine Cd responsive genes, which included a metal transporter, putative transcription factor and an NHL repeat membrane-spanning protein. Additional candidate genes located in the QTL intervals included a glutathione-S-transferase and putative homolog of a glutamine cysteine ligase.

Several QTL studies on rice (*Oryza sativa* L.) have been conducted to determine the number of genes involved in metal accumulation and tolerance. Three putative QTLs involved in Cd accumulation have been found on chromosomes 3, 6 and 8 (Ishikawa *et al.*, 2010; Ishikawa *et al.*,2005). Ueno *et al.* (2009) also identified another major QTL for Cd accumulation in *O. sativa* that was mapped on the short arm of chromosome 7. QTLs for the translocation of Cd from roots to sink regions were reported in *O. sativa* (Xu *et al.*, 2012; Tezuka *et al.*, 2010). Tezuka *et al.* (2010) revealed a major QTL (*qCdT7*), mapped to chromosome 7, which controlled the translocation of Cd from roots to shoots. This QTL explained 88% of the phenotypic variation indicating that low Cd accumulation was a dominant trait. Dufey *et al.* (2009), using recombinant inbred lines, identified in*O. sativa* 24 putative QTLs involved in Fe tolerance which were mapped to chromosomes 1, 2, 3, 4, 7 and 11. In addition, two QTLs, located on chromosomes 2 and 3, were involved in As concentration in shoots and in roots respectively.

In durum wheat (*Triticum durum*, L.), Cd accumulation is controlled by a major gene named *Cdu1*and localized on chromosome 5BL (Knox *et al.*, 2009; Clarke *et al.*, 1997). Further, Ci *et al.* (2012) characterized 26 QTLs involved in Cd tolerance and accumulation in *T. aestivum*, where 16 were involved in Cd stress control, 8 for Cd tolerance and 2 for Cd accumulation in roots. In *A. sativa* L., a single QTL for Cd accumulation in grain has been reported (Tanhuanpää *et al.*, 2007).

In wheat (*T. aestivum* L.), Mayowa and Miller (1991) reported QTLs involved in Cu tolerance and accumulation that were mapped to chromosomes 5A, 4D, 7A, 7B, 7D. Ganeva *et al.* (2003) also characterized QTLs for *T. aestivum* on chromosomes 1A, 1D, 3A, 3B, 4A and 7D. Bálint *et al.*(2003) identified QTLs associated with Cu tolerance located on *T. aestivum* chromosomes 3D, 5A, 5B, 5D, 6B and 7D. In addition, Bálint *et al.* (2007) also determined QTLs for Cu tolerance in *T. aestivum*. They reported one major QTL for Cu tolerance on chromosome 5D and minor QTLs on chromosomes 1A, 2D, 4A, 5B and 7D. A QTL affecting shoot Cu content under Cu stress conditions was mapped on chromosome 1BL and an additional QTL for Cu accumulation was found on chromosome 5AL. The role of these genes located on various chromosomes in these different studies suggests that Cu tolerance is a polygenic character, as well as the possibility of different gene expressions against distinct toxic Cu concentrations in different populations. The accumulation of Cu in the shoots is affected by different QTLs, suggesting a strong metal-specific uptake and/or translocation. Bálint *et al.* (2007) reported a negative correlation between Cu tolerance and accumulation in the shoot indicating that the key tolerance mechanism in wheat could be the restriction of Cu uptake in the roots or the reduced translocation from root to shoot.

Categories of Plants Growing on Metal Contaminated Soils

Baker and Walker (1990) categorized plants into three groups according to their strategy for coping with metal toxicity in soil; metal excluders, indicators and accumulators/ hyperaccumulators.

Excluders

The metal excluder strategy consist in limiting the amount of metals translocated from roots to shoots thus maintaining low levels of metal concentration in their aerial parts. Large amounts of metals in the roots of excluder species have been reported (Baker and Walker 1990). Examples of excluder species include *Oenothera biennis, Commelina communis, Silene maritime, Agrostis stolonifera* L., woody plants such as *Salix, Populus* and *Pinus radiata* (Maestri *et al.*, 2010; Wei *et al.*, 2005).

Accumulators/Hyperaccumulators

Metal accumulators/hyperaccumulators are plants that can concentrate metals in their above-ground tissues to levels that exceed those in the soil or also to those in the non accumulating species found growing nearby with concentrations up to 100 times more than non hyperaccumulators (Salt *et al.*,1998). Accumulators/hyperaccumulators growing on metal contaminated environments can naturally accumulate higher levels of heavy metals in their shoots than in their roots (Kachout *et al.*, 2009). Some plants can accumulate only a specific metal while others can accumulate multiple metals ((Mganga *et al.*, 2011; Almås *et al.*, 2009).

Presently, at least 45 plant families comprising more than 400 species have been found to accumulate metals in their harvestable tissues, and the majority of them belong to the *Brassicaceae* family (Pal and Rai 2010). The best known genera from this family are *Alyssum* and *Thlaspi*. *Thlaspi* species can accumulate more than 3% of their shoots in Zn, 0.5% in Pb and 0.1% in Cd. *A. halleri* can also accumulate more than 1% of its above-ground biomass in Cd and Zn and *Alyssum* species can accumulate over 1% Ni in their harvestable parts (Di Baccio *et al.*, 2011). There are variations among family, species and populations in the ability to accumulate metals. For example, *Arabidipsis halleri* can accumulate Cd and Zn in their harvestable parts where as *A. thaliana* is known to be a metal excluder and restricts metals in the roots. *Betula spp.* can accumulate Zn, while other trees species of the same family (*Carpinus* and *Corylus*) are unable to do so (Ernst, 2006; Ernst, 2004).

Indicators

Like accumulators, metal indicators accumulate metals in their aerial tissue, but the metal levels in the above ground tissue of these plants usually reflect the metal concentration in the surrounding environment (Baker and Walker 1990). If these plants continue to uptake metals, they will eventually die-off. These plants are of biological and ecological importance since they are pollution indicators and also, like accumulators, they absorb pollutants (Mganga *et al.*, 2011).

Determination of Excluders, Indicators and Hyperaccumulators Plants

A plant is classified as a hyperaccumulator when it meets four criteria including; a) when the level of heavy metal in the shoot divided by level of heavy metal in the root is greater than 1 (shoot/root quotient > 1); b) when the level of heavy metal in the shoot divided by total level of heavy metal in the soil is greater than 1 (extraction coefficient > 1) (Rotkittikhun *et al.*, 2006; Harrison and Chirgawi 1989); c) when the plant takes up between 10 – 500 times more heavy metals than normal plants (uncontaminated plants - control plants) (Fifield and Haines 2000; Allen,1989); and d) more than 100mg/kg of cadmium, 1000g/kg of copper, lead, nickel, chromium; or more than 10000mg/kg of zinc (Mganga *et al.*, 2011; Ernst, 2006; Brooks, 1998). An excluder is a plant that has high levels of heavy metals in the roots but with shoot/root quotients less than 1 (Boularbah *et al.*, 2006). Finally,Baker and Walker (1990) classified a plant as an indicator when the levels of heavy metals within their tissues reflect those in the surrounding soil.

Physiological Mechanisms of Metal Resistance

Resistant plants are able to grow on metal contaminated soil due to avoidance and/or tolerance strategies. Plant resistance to high levels of heavy metals in soils can result from either reduced uptake or once taken up, metals have to be transformed into a physiologically tolerable form.

Restriction of Metal Uptake

The plasma membrane is the first structure of living cells exposed to heavy metals. The membrane functions as a barrier for the movement of heavy metals into cytoplasm. The restriction of metals at the plasma membrane limits the uptake and accumulation of metals by preventing their entry into the cytoplasm. This can be done by changing the ion binding capacity of the cell wall and/or decreasing the uptake of metal ions through modified ion channels, and/or by removing metals from cells with active efflux pumps and/or with root with root exudates (Tong *et al.*, 2004).

The Cell Wall

The cell wall and membrane interface could be a site of metal tolerance since a significant amount of metals has been reported to be accumulated there. Divalent and trivalent metal cations can bind plant cell walls because of the presence of functional groups such as –COOH, -OH and –SH. Pectins are polymers that contain carboxyl groups which enable the binding of divalent and trivalent heavy metals ions. In enriched heavy metal environments, some plants will increase the capacity of their cell wall to bind metals by increasing polysaccharides, such as pectins (Colzi *et al.*, 2011; Pelloux *et al.*,2007). Konno *et al.* (2010; 2005) showed that the pectin in root cell walls was important in binding Cu in the fern, *Lygodium japonicum,* and in the moss, *Scopelophila cataractae*. The cell wall of*Minuartia verna sp. hercynica* growing on heavy metal contaminated medieval mine dumps has been found to have high concentrations of Fe, Cu, Zn and Pb (Solanki and Dhankhar 2011;Neumann *et al.*, 1997). On the other hand, Colzi *et al.* (2012) found that a copper tolerant *Sileneparadoxa* population restricted the accumulation of Cu in roots, when exposed to high Cu, by decreasing their pectin concentration in the cell wall and increasing pectin methylation thus preventing the binding of Cu.

Root Exudates

Resistant plants can also restrict the entry of metals by immobilizing them in the rhizosphere with root exudates outside the plasma membrane (Colzi *et al.*, 2011). This has been reported in *T. aestivum* where the exudation of phytochelatins, citrate and malate may be responsible for Cu exclusion mechanisms in non accumulators (Yang *et al.*, 2005b; Bálint *et al.*, 2007). Hall (2002) also proposed a mechanism for Ni exclusion in plants involving Ni-chelating exudates which include histidine and citrate. In non hyperaccumulator plants, these Ni chelators accumulate in their root exudates which, in turn decreases Ni uptake. The copper exclusion could be due to its chelation with citrate and malate exudates in the rhizosphere of wheat roots. The restriction of Cu uptake in wheat by the efflux of these organic acids has been previously documented by Nian *et al.* (2002).

Chelation

The phytotoxic effect of free metal ions can be eliminated by their chelation by specific high-affinity ligands (Yong and Ma 2002). The chelation of metals allows for the restriction of metal uptake, the uptake of metal ions, sequestration and compartmentation, as well as xylem loading and transport within the plant. Baker *et al.* (2000) categorized these ligands according to the characteristic electron donor centers, which include sulfur donor ligands, oxygen donor ligands and nitrogen donor ligands.

Oxygen Donor Ligands

Organic acids such as malate, aconitate, malonate, oxalate, tartrate and citrate are involved in metal uptake restriction and dexotification in plants. These carboxylic acid anions form complexes with divalent and trivalent metal ions with high stability. They are involved in the restriction of metal entry into the cell, metal exclusion in the root cells, accumulation and transport within the plants. In wheat (*T. aestivum*), citrate and malate formed complexes with Cu in order to immobilize this metal in the rhizosphere thus preventing its entry into the cell (Yong and Ma 2002). Citrate was also involved in the hyperaccumulation of Ni in 17 New Caledonian plants and the amount of citrate produced was highly correlated with the accumulation of Ni (Lee *et al.,* 1977). The accumulation of Zn in some plants is facilitated by the transport of malate-Zn complexes. Upon the Zn ions uptake into the cytoplasm, they are bound to malate, which serves as a carrier to transport the Zn ions to the vacuole. Once there, the Zn ions are complexed by a terminal acceptor and released from malate. The malate is then able to return to the cytoplasm and transport additional Zn ions to the vacuole (Yong and Ma 2002). Still and Williams (1980) reported that the transport of free Ni ions to root cells via membrane is restricted. However, when Ni is bound to organic compounds such as citric and malic acids, it can be transported across the plasma membrane (Yong and Ma 2002). In *Zea mays*, the production of organic acid is influenced by external aluminium ion concentration (Pintro *et al.,*1997). Also, in manganese tolerant *T. aestivum* cultivars, the production of malic, citric or aconitic acid was not induced when exposed to this metal but for the manganese sensitive cultivars, the organic acids concentration slightly increased (Burke *et al.,* 1990).

Nitrogen Donor Ligands

This group consists of amino acids and their derivatives which have relatively high affinity for specific metals. Krämer *et al.,* (1996) revealed histidine to be involved in the Ni tolerance and translocation of the hyperaccumulator plant *Alyssum lesbiacum*. The majority of Zn in roots of the Zn hyperaccumulator, *T. caerulescens*, was complexed with histidine (Salt *et al.,* 1999). Studies have also shown histidine to be involved in the restriction of metal uptake. For example, plants chelate Ni with histidine in the rhizosphere which prevents the uptake of this metal (Wenzel *et al.*, 2003).

Sulfur Donor Ligands

In plants, sulfur donor ligands are composed of two classes of metal chelating ligands which are phytochelatins (PCs) and metallothioneins (MTs). Phytochelatins are small metal binding peptides synthesized from the tripeptide glutathione (γ-Glu-Cys)$_{2\text{-}11}$-Gly) (Solanki and Dhankhar 2011; Hall, 2002). Since there is a γ-carboxamide linkage between glutamate and cysteine, PCs are not synthesized by translation of mRNA, but rather it is a product of an enzymatic reaction involving the enzyme PC synthase (Yong and Ma 2002). The production of PCs is positively correlated with metal accumulation in plant tissues (Pal and Rai 2010). PCs are produced in cells immediately after heavy metal exposure, including Cd, Pb, Zn, Ag, Hg, As and Cu as seen in *Rubia tinctorum* (Maitani *et al.*,1996). PC production can be induced in roots, shoots, and leaves as observed in *Sedum alfredii*when exposed to Cd (Pal and Rai 2010).

Several research groups concurrently and independently cloned and characterized genes encoding PC synthase. These genes were isolated from *Arabidopsis thaliana, Schizosaccharomyces pombe,* and*T. aestivum*, and were designated *AtPSC1, SpPCS,* and *TaPCS1*, respectively. They encoded 50-55kDa sequences with 40-50% similarity. The polypeptides were found to be active in the synthesis of PCs from glutathione (GSH) (Yong and Ma 2002). In cultured *Silene cucubalis* cells, the presence of heavy metals, such as Cd, Cu, Zn, Ag, Hg and Pb, induce the synthesis of PCs by PC synthase from the GSH like substance (Pal and Rai 2010). Gaudet *et al.* (2011) did a comparative analysis of two *Populus nigra* genotypes from contrasting environments. They determined that both genotypes responded differently to Cd stress. The southern genotype (Poli) was more tolerant than the northern genotype (58-861). This variation was due to different adaptation strategies to Cd stress. The thiol and PC content, which was associated with the *glutathione S-transferase* gene, was higher in the southern genotype as compared to the northern genotype, which under Cd stress, revealed differences in the use of phytochelatin pathway that might be related to the variation in their Cd tolerance.

The second class of sulfur donor ligands are metallotioneins (MTs). They are low molecular weight (4-14kDa), cysteine-rich, metal-binding proteins found in a wide range of organisms (animals, plants, eukaryotic microorganisms, and prokaryotes) (Huang and Wang 2010). Unlike PCs, they are encoded by structural genes (Yong and Ma 2002). They play essential roles in a variety of organisms including Cu, Cd and Hg detoxification by sequestration (Palmiter, 1998; Ecker *et al.*, 1989), Zn homeostasis (Coyle *et al.*, 2002) and also scavenging of reactive oxygen species (Wong *et al.*, 2004). MTs have been divided into two classes based on their cysteine residue arrangements. Class I MTs are widespread in vertebrates and are composed of 20 highly conserved cysteine residues based on mammalian MTs. Class II MTs have slightly flexible cysteine arrangements and are found in plants, fungi and invertebrates. A third class includes phytochelatins (Chaturvedi *et al.*, 2012). Based on the position and allocation of cysteine residues, class II plant MTs are additionally divided into four types (Cobbett and Goldsbrough 2002). Type 1 plant MT genes have been more highly expressed in roots compared to leaves while the reverse is observed for the expression of type 2 plant MT genes. Type 3 MT genes are highly expressed in ripening fruits or in leaves while the expression of type 4 plant MT gene is restricted to developing seeds (Sekhar *et al.*, 2011; Cobbett and Goldsbrough 2002).

The expression of MT genes in plants subjected to metal stress has been studied. *AtMT1* and *AtMT2*genes showed increased expression levels when *Arabidopsis* plants were exposed to high levels of Cu and Cd (Sekhar *et al.*, 2011). Van Hoof *et al.* (2001) reported that the copper tolerant *S. vulgaris*individuals showed higher *SvMT2b* expression in roots and shoots when exposed to high concentrations of copper compared to the copper sensitive plants. Huang and Wang (2009) reported an increase in BgMT2 mRNA expression in large-leafed mangrove plants (*Bruguiera gymnorrhiza*) when exposed to Zn, Cu or Pb. Similar results were described by Gonzalez-Mendoza *et al.* (2007) in black mangrove *(Avicennia germinans*) seedlings exposed to Cd or Cu, showing a significant increase in AvMT2. High levels of the *CcMT1* transcripts were also observed in pigeon pea (*Cajanus cajan* L.) exposed to Cd and Cu (Sekhar *et al.*, 2011).

In general, there are variations between species in the expression of MTs to various metals. The up and down regulation of MTs in response to metal stress is largely unknown in plants. The MT gene expression was shown to be strongly induced by Cu, Cd, Pb and Zn (Huang and Wang 2009;Gonzalez-Mendoza *et al.*, 2007; van Hoof *et al.*, 2001). MT gene expression is also influenced by other abiotic stressors including absciscic acid (ABA), drought, salinity, heat, cold light, wounding and senescence (Sekhar *et al.*, 2011).

Mechanisms Involved in Internal Metal Tolerance

Metal Uptake

The uptake of metal from soil into roots is dependent on the bioavailabilty of the metal, as well as its mobility in the rhizosphere (Maestri *et al.*, 2010). The bioavailability of various metals greatly varies. No correlation exists between the metal content in soils and in plants (Clemens, 2006). The bioavailability of metals in the rhizosphere is affected by the chemical environment. For example, in*T. caerulescens,* the chemical form of nitrogen influences the plants ability to uptake Cd and Zn (Maestri *et al.*, 2010; Xie *et al.*, 2009). Metals present in the rhizosphere of hyperaccumulators are more bioavailable than for those of non hyperaccumulators. Plants can render metals mobile in their rhizosphere by excreting root exudates, such as organic acids and phytosiderophores and by acidification with protons (Maestri *et al.*, 2010; Marschner,1995). Bacteria in the soil also affect metal mobility and availability by lowering the pH, producing hormones, organic acids, antifungals, antibiotics and metal chelators which all enhance the root growth (Maestri *et al.*, 2010; Wenzel *et al.*,2003). Higher amounts of bacteria were found in the rhizosphere of hyperaccumulators. Microorganisms found in the rhizosphere were linked to an increased uptake of Cd, Z, and Pb in*Sedum alfredii* and an enhanced root growth (Maestri *et al.*, 2010; Xiong *et al.*, 2008).

Metal Uptake across the Plasma Membrane

The uptake of heavy metals in plants is mediated by a group of metal transporter families which consists of iron-responsive transport proteins (ZIP-IRT), the heavy metal-transporting P_{1B}-type subfamily of

P-type ATPases, the natural resistance associated macrophage proteins (NRAMP) and the cation diffusion facilitators (CDF) (Baxter *et al.*, 2003). Transporters were originally identified for Fe^{2+} or Zn^{2+} homeostasis, but it was demonstrated that most transporters of essential metal ions can also carry non-essential metals, such as Cd (Zhou *et al.*, 2012). The uptake of non essential metals may be the result of their close chemical characteristics or metal ion size to essential metals. Some metal transporters, present in the plasma membrane of root cells, exhibit low substrate specificity which can lead to the accumulation of other metals in plants (Schaaf *et al.*, 2006). For example, the non-functional metal Cd can be taken up via a Ca^{2+} transporter (Perfus-Barbeoch *et al.*, 2002) or also via the Fe^{2+} transporter IRT1 (Korshunova *et al.*, 1999). Plant tolerance to metal stress can be achieved with the modification of these transporter activities (Zhou *et al.*, 2012). Plants can prevent the uptake of certain metals by down-regulating the expression of such transporters, as observed in *S. vulgaris,* where the tolerant plants restrict the uptake of Cu by the down-regulation of Cu-transporters (Assunçãno *et al.*, 2003; Harmens *et al.*, 1993). Since Fe and Ni belong to the group of transient metals and have similar chemical properties, Fe deficiency may be the result of Ni phytotoxicity. Ni competes with Fe in physiological and biochemical processes, and in turn roots, can uptake Ni by Fe transporters (Pandey and Sharma 2002).

Increased Zn uptake is driven by an overexpression of members of the ZIP family of transporters. Under Zn deficiency conditions, many members of the ZIP transporter family are overexpressed in non hyperaccumulator species, while in hyperaccumulators, they are independently expressed regardless of Zn supply (Verbruggen *et al.*, 2009). Nishida *et al.* (2011) and Schaaf *et al.* (2006) showed that *A. thaliana* can increase the uptake of Ni in roots when Fe levels are low by the Iron-Regulated Transporter 1 (AtIRT1; member of Zrt/IRT-like *ZIP* family of transporters). AtIRT1 has a wide specificity for divalent heavy metals including Ni, Zn, Mg, Co and Cd and mediates the accumulation of such metals under Fe-deficient conditions. Nakanishi *et al.* (2006) reported that Cd was uptaken in yeast by two *O. sativa* Fe^{2+} transporters, *OsIRT1* (Iron-Regulated Transporter 1) and*OsIRT2.*

The uptake of Ni of some Ni hyperaccumulator accessions of *Thlaspi goesingense, Thlaspi japonicum* and *T. caerulescens* has been reported to be inhibited in the presence of Zn. This demonstrated that Ni entered the cell via Zn uptake transporters, specifically the TcZNT1 transporter (Assunçãno *et al.*, 2008). In Zn deficiency conditions, the expression of AtZIP4, the orthologue of TcZNT1 in *A. thaliana,* can be induced but when additional Ni was added, the expression was repressed. This suggested that Zn and Ni competed for their uptake via AtZIP4/TcZNT1 transporters (Hassan and Aarts 2011). In addition, in presence of high Zn concentration, the expression of *ZNT1*was higher in Zn hyperaccumulator *T. caerulescens* roots than in the non hyperaccumulator *Thlaspi arvense* suggesting its involvement in the hyperaccumulator phenotype (Hassinen *et al.*, 2007;Assunçãno *et al.*, 2001; Assunçãno *et al.*, 2001; Pence *et al.*, 2000). Milner *et al.* (2012) also determined that NcZNT1, isolated from *T. caerulescens*, played a role in Zn uptake from the soil which was based on its high expression in root.

Heavy metal-transporting P_{1B}-type transporters are also involved in metal-ion homeostasis and tolerance in plants by transporting essential and non essential heavy metals such as Cu, Zn, Cd, Pb across cell membrane. Transporters located at the plasma membrane function as efflux pumps by removing toxic metals form cytoplasm. They have also been found in membranes of intracellular organelles for compartmentalization of metals for sequestration in vacuoles, golgi or endoplasmic reticulum (Yang *et al.*, 2005b). These ion pumps transport ions across a membrane by hydrolysing ATP (Benitez *et al.*, 2012). Eight P_{1B}-ATPases, AtHMA1–AtHMA8, have been reported in*Arabidopsis* (Baxter *et al.*, 2003). AtHMA1, 2, 3, and 4 showed high similarity with $Zn^{2+}/Co^{2+}/Cd^{2+}/Pb^{2+}$ ATPases previously characterized in prokaryotes (Axelsen and Palmgren 2001). The AtHMA4 was located at the plasma membrane. The ectopic expression of AtHMA4 improved the growth of roots in the presence of toxic Zn, Cd and Co concentrations (Yang *et al.*,2005b). The heterologous expression of AtHMA4 enhanced Cd tolerance in yeast (Mills *et al.*,2003).

In addition, the gene *Nramp* encodes for another divalent metal transporter located at the plasma membrane. This transporter also removes toxic metals from the cytosol by efflux pumping. It has been reported to be expressed in roots of *Arabidopsis* and *O. sativa* (*OsNramp1*- expressed in rice roots where as *OsNramp2* is expressed in leaves and *OsNramp3* is expressed in both tissues). The*OsNramp1* gene was found to be involved in the uptake of Mn, while the *Nramp* genes in*Arabidopsis* and rice were involved in the uptake of Cd, and other divalent metals (Yang *et al.*,2005b). The AtNRAMP1,

3, and 4 showed uptake of Cd^{2+} when they were expressed in the yeast*Saccharomyces cerevisiae*. In addition, Cd^{2+} hypersensitivity was observed in *A. thaliana* when AtNRAMP3 was overexpressed. This transporter was located in the vacuolar membrane where it is involved in the mobilization of metals from the vacuole (Clemens, 2006).

In bacteria and in some eukaryotes, Zn, Co and Cd are transported by the CDF transport proteins. Within the *Arabidopsis* genome, there are 12 nucleotide sequences that are predicted to encode members of CDF transporter family. However, these transporters might be involved in cation efflux out of the cytoplasm, by pumping ions out of the cytoplasm to the exterior of the cell or into intracellular compartments such as the vacuole (Yang *et al.*, 2005b).

Plants can make metal ions more available for uptake by acidifying the rhizosphere and pumping protons via plasma membrane-localized proton pumps; and also by exuding low molecular weight (LMW) compounds that act as metal chelators (Clemens, 2006). The secretion of organic acids can render heavy metals mobile and enhance their absorption by plant roots. Krishnamurti *et al.* (1997) reported that when Cd was complexed with organic acids, it was readily available for transport across the membrane, while free Cd ions were restricted for uptake. Cieśliński *et al.* (1998) revealed a higher acetic acid and succinate in the rhizosphere of the *T aestivum* (Kyle) Cd accumulating genotype compared to the non accumulating (Arcola) wheat genotype. The Zn/Cd hyperaccumulating*Sedum alfredii* was able to extract high levels of Zn and Pb from its contaminated environment because of the release of root exudates (Li *et al.*, 2005). In *Alyssum*, the Ni transport and accumulation was enhanced by secretion of histidine in the rhizosphere (Krämer *et al.*, 1996).

Sequestration/Compartmentation

Some metal tolerant plants can accumulate large amounts of metals within the cell without exhibiting toxicity symptoms (Entry *et al.*, 1999). These plants are able to store the surplus of accumulated metals where no sensitive metabolic activities occur such as organs or subcellular compartments (Ernst, 2006). This avoidance of metal poisoning involves the intracellular sequestration and apoplastic or vacuolar compartmentation of the toxic metal ions (Liu *et al.*, 2007). Compartmentation of metals can also be found in the cells central vacuole. This was observed in the Zn resistant*Deschampsia cespitosa* where the excess Zn ion was removed from the cytoplasm and actively pumped into the vacuoles of root cells where as Zn sensitive plants had a much lower capacity to do so (Brookes *et al.*, 1981).

Schaaf *et al.* (2006) determined that the transporter AtIREG2, located at the tonoplast, was involved in Ni detoxification in roots. AtREG2, confined to roots, prevents heavy metal translocation to shoots restricting metals to roots. This transporter counterbalances the low substrate specificity of transporter AtIRT1 and other iron transporters in iron deficient root cells. The AtIREG2 transporter, found in *A. thaliana*, was involved in the detoxification of Ni in roots under Fe deficiency conditions at pH 5 (Schaaf *et al.*, 2006). The *T. caerulescens ZTP1* gene was involved in the intracellular sequestration of Zn. The expression of the *ZTP1* gene was higher in the roots and shoots of the Zn tolerant *T. caerulescens* compared to the non tolerant plant (Assunçãno *et al.*, 2001).

Members of the CDF protein play a role in tolerance to various metals including Cd, Co, Mn, Ni and Zn by their sequestration into vacuoles (Montanini *et al.*, 2007). Increased Zn tolerance and accumulation was reported in non accumulator *A. thaliana* when *AtMTP1*, *PtdMTP1*, *AtMTP3* and *TgMTP1* (members of the CDF family) were ectopically or heterologously expressed. This suggested that the function of these proteins was the creation of a sink of Zn in the vacuole of plant cells in instances of high intracellular Zn levels or as buffer in Zn deficiency situations (Hassan and Aarts 2011).

Phytochelatins are also thought to be involved in the restriction of metals to the roots (Zenk, 1996). When *Nicotiana tubacum* seedlings were exposed to excess Cd, the level of phytochelatin increased (Vogelilange and Wagner 1990). The metal-phytochelatin complexes are formed when plants are exposed to high heavy metal concentrations. They are then sequestered into vacuoles for detoxification. A group of organic solute transporters actively transport phytochelatin-metal complexes into the plant's vacuole (Solanki and Dhankhar 2011; Salt and Rauser 1995). In the presence of excess Cu and Cd, phytochelatins form complexes with these metals in *Zea mays* and in turn reduce the root to shoot translocation (Galli *et al.*, 1996). The synthesis of phytochelatins is catalyzed by the enzyme phytochelatin synthase (PCS), a constitutive enzyme which requires post-translational activation by heavy metals and/or metalloids that include Cd, Ag, Pb, Cu, Hg, Zn,

Sn, As and Au (Solanki and Dhankhar 2011). Martínez *et al.* (2006) reported that the expression of a PCS gene isolated from *T. aestivum* improved the accumulation of Cd, Pb and Cu in *Nicotianaglauca*. The elevation of phytochelatin concentration in roots might reduce the root to shoot transport required for accumulation in shoots.

Root to Shoot Translocation

The translocation of metals to the aerial biomass can be an important biochemical process used by plants to remediate polluted areas. In some plants, the mobilization of metals from their roots to their above aerial organs can minimize the damage that could be exerted by these heavy metals on the root physiology and biochemistry (Zacchini *et al.*, 2009). Excluders prevent or limit the translocation of toxic metals or essential metals from roots to shoots. On the other hand, accumulators/hyperaccumulators translocate metals from roots to shoots via the xylem with the transpiration stream. This is accomplished by increasing the uptake of metals in roots, and by reducing the sequestration of metals in the root.

The chelation of metals with ligands, such as organic acids, amino acids and thiols facilitates the movements of heavy metals from roots to shoots (Zacchini *et al.*, 2009). The xylem cell wall has a high cation exchange capability, thus the movement of metal cations is severely retarded when the metals are not chelated by ligands. Organic acids are involved in the translocation of Cd in the species*Brassica juncea* (Salt *et al.*, 1995).

The chelation of Ni to histidine is involved in the long distance translocation of Ni in the hyperaccumulator *A. lesbiacum*, where a 36-fold increase was reported in the histidine content of the xylem sap upon exposure to nickel (Solanki and Dhankhar 2011; Krämer *et al.*, 1996). Richau *et al.*(2009) found that the Ni hyperaccumulator, *T. caerulescens,* had a higher free histidine concentration in roots compared to the non Ni hyperaccumulator *T. arvense*. Also, *T. caerulescens* had less Ni in root vacuoles than *T. arvense* because the histidine-Ni complexes were much less taken up by vacuoles than free Ni ions. Therefore, an increase in free histidine in roots inhibited the vacuolar sequestration of His-Ni in *T. caerulescens* compared to free Ni in *T. arvense* and also had enhanced histidine-mediated Ni xylem loading. The elevated free histidine in root cells appears to be involved in reduced vacuolar sequestration and enhanced xylem loading of Ni (Richau and Schat 2009). This was also the case for Zn and Cd for this hyperaccumulating species (Hassan and Aarts 2011). An increase in Ni accumulation was also observed in the Ni hyperaccumulator *Sebertia acuminata* where, when chelated to citrate, Ni was able to translocate to the shoot. In the absence of citrate, Ni was no longer accumulated in the aerial tissues (Lee *et al.,* 1977).

The chelation of metals with nicotianamine (NA) also contributes to improved tolerance. Nicotianamine can chelate and transport divalent Ni, Cu and Zn (Takahashi *et al.,* 2003; Pich *et al.,*2001; Ling *et al.,* 1999). The nicotianamine synthase (NAS) enzyme is responsible for the synthesis of NA by trimerization of S-adenosylmethionine (Shojima *et al.,* 1990). When exposed to high levels of Zn, Cd, and/or Ni, all four *NAS* genes were highly expressed in *T. caerulescens* compared to non hyperaccumulator *A. thaliana* (van de Mortel *et al.,* 2006). In the presence of elevated Mn, Zn, Fe and Cu concentrations, Kim *et al.* (2005) reported an increased expression of the *NAS* gene, as well as NA levels for *A. thaliana* and *N. tubacum*. In addition, Pianelli *et al.* (2005) showed that the over-expression of the *T. caerulescens NAS3* gene in the Ni excluder *A. thaliana* resulted in improved Ni tolerance and Ni accumulation in their aerial organs. An increase of Fe, Zn and Cu accumulation in*O. sativa* was associated with an overexpression of the *NAS3* gene (Hassan and Aarts 2011; Kawachi*et al.,* 2009).

Visioli *et al.* (2010) also showed that metallothioneins may be involved in the translocation of Ni in *T. caerulescens.* An increase in MT-1B in the individuals from the metal contaminated environment was observed when metallicolous *T. caerulescens* and non-metallicolous *T. caerulescens* individuals were grown in presence of high Ni concentrations, compared to non contaminated site. Additionally, Visioli*et al.* (2012) analyzed four *T. caerulescens* sub-population (MP1 to MP4) for their ability to accumulate and tolerate Ni. In four sub-populations analyzed, MP2p translocated the highest amount of Ni to the shoots. This sub-population also had the highest level of putative metallothionein protein (MT4C). Constitutively higher expressions of other MTs are also seen in the hyperaccumulators *A. halleri, S. paradoxa* and *S. vulgaris*.

Transporters are not only involved in the uptake of metals from the soil, but also in their transport out of the vacuole. These mobilized metals can then be translocated to aerial tissue. Visioli *et al.* (2012) subsequently found for sub-population MP2p, which exhibited the highest level of Ni translocation of the four sub-populations analyzed, significantly higher levels of the ABC27 transporter. This transporter is part of the ABC family of transporters which are involved in removing metals from the cytoplasm by pumping outside the cell wall, metals sequestered in vacuoles and other subcellular compartments (Visioli *et al.*, 2012; Martinoia *et al.*, 2002; Sanchez-Fernandez *et al.*,2001). Hassinen *et al.* (2007) showed that the AtMRP10 homolog, also part of the ABC family of transporters, had different expression in roots of two *T. caerulescens* populations with contrasting Zn tolerance and accumulation. In addition, the AtNramp3 transporter was also involved in the mobilization of vacuolar Cd back into the cytosol. This was oberved when *AtNramp3* was overexpressed in *A. thaliana*. AtNramp3 was further hypothesized to play a role in the mobilization of Fe, Mn, and Zn in the vacuole (Clemens, 2006).

The passage of metal ions and/or metal ligand complexes from the cytosol of root cells into the vascular tissue requires their transport across the cell membrane. Transporters involved in this activity are the heavy metal transporting P-type ATPases (HMAs) (Clemens, 2006). The AtHMA2 and 4 are involved in translocation of Zn in *A. thaliana*. Stunted growth and chlorosis resulted in the*hma2hma4* double mutant from inadequate Zn supply to the leaves. The two genes were expressed in vascular tissue which indicates their hypothesized function in xylem loading (Hussain *et al.*, 2004). The AtHMA4 transporter was also involved in the transport of Cd^{2+} ions (Clemens, 2006). In *T. caerulescens*, the P-type ATPase, TcHMA4, was also involved in the translocation of Zn. When Zn and Cd levels were elevated or when Zn is deficient, the expression of *TcHMA4* was induced in the roots. This transporter was involved in the xylem loading of Zn in plant roots (Hassinen *et al.*, 2007;Papoyan and Kochian 2004). Milner *et al.* (2012) also determined that NcZNT1 in *T. caerulescens*was not only involved in Zn uptake from the soil but also could be involved in the long distance transport of Zn from root to shoot via the xylem.

Metal Storage

Metals have to undergo a xylem unloading process prior to their distribution and their detoxification in the shoot and their redistribution via the phloem (Schmidke and Stephan 1995). Once unloaded, the metals are either taken up into surrounding cells and are symplastically transported through the leaf tissues or they are apoplastically distributed over the leaf (Hassan and Aarts 2011; Marschner, 1995). NA is important in the chelation of metals for their symplastic transport through the leaf. This occurs through the Yellow Stripe Like proteins (YLS) (Hassan and Aarts 2011; DiDonato *et al.*, 2004). In the hyperaccumulator *T. caerulescens*, three *YSL* genes (*TcYLS3*, *TcYSL5* and *TcYSL7*) were highly expressed in shoots around vascular tissues. This high level of expression was not observed in the excluder plant *A. thaliana* orthologues (Hassan and Aarts 2011; Gendre *et al.*, 2007). For the*TcYSL3*, it was suggested that its function was to unload Ni-NA complexes from the xylem into leaf cells and to distribute it to storage cells. Using yeast complementation and uptake measurement studies, it was determined that TcYSL3 was also a Fe/Ni-NA influx transporter. Considering that YSL proteins have a role in the transport of Fe-NA complexes, it was proposed that they might also be involved in the hyperaccumulation of Fe-NA in some plants (Hassan and Aarts 2011; Curie *et al.*,2009).

The sequestration of excess essential and non essential metals is localized in various parts of the aerial tissue, such as trichomes, leaf epidermal cell vacuole and mesophyll vacuole. Broadhurst *et al.*(2004) grew five *Alyssum* hyperaccumulator species/ecotypes on Ni-enriched soil and determined that the majority of hyperaccumulated Ni was stored in either leaf epidermal cell vacuoles or in the basal section of stellate trichomes. They also found that the metal concentration in the basal part of the trichome was 15% to 20% of dry weight. This was among the highest metal concentrations reported in healthy vascular plant tissues. In *A. halleri*, the majority of Zn ions were stored in the vacuoles of mesophyll cells, while for *T. caerulescens*, most Zn ions were located in the vacuoles of epidermal cells (Verbruggen *et al.*, 2009). The transport of metals through the phloem sap is less documented. The sole molecule identified as a phloem metal transporter is nicotianamine which is involved in the transport of Fe, Cu, Zn and Mn (Stephan *et al.*, 1994).

Antioxidative Defence Involved in Metal Tolerance

In environments, where metals are present in toxic levels, the elevated activities of antioxidant enzymes and non-enzymatic constituents are important in the plant tolerance to stress. Metal tolerance may be enhanced by the plant's antioxidant resistant mechanisms. There is an indication that the alleviation of oxidative damage and increased resistance to stresses in the environment is often correlated with an effective antioxidative system. The minimization of damage due to oxidative stress is a universal feature of plants defense responses (Kachout *et al.*, 2009). The detrimental effect of heavy metals in plants is due to the production of ROS and induction of oxidative stress. Oxidative stress is expressed by the increase levels of reactive oxygen species such as singlet oxygen ($^1O_2$), superoxide radical (O^{-2}), hydrogen peroxide (H_2O_2) and hydroxyl radical (OH^-) (Salin, 1988). ROS are strong oxidizing agents that lead to oxidative damage to biomolecules, for instance lipids and proteins and can eventually result in cell death (Gunes *et al.*, 2006). It is shown that plant tolerance to metals is correlated with a rise in antioxidants and activity of radical scavenging enzymes (Kachout*et al.*, 2009). Plants respond to oxidative stress by activating antioxidative defence mechanisms which involve enzymatic and non-enzymatic antioxidants. The enzymatic components include superoxide dismutase (SOD), catalase (CAT), ascorbate peroxidase (APX) and enzymes of ascorbate glutathione cycle whereas the non-enzymatic antioxidants include ascorbate and glutathione and atocoperol (Solanki and Dhankhar 2011; Kachout *et al.*, 2009). These antioxidants are responsible for elimination and destruction of the reactive oxygen species (Solanki and Dhankhar 2011).

Oxidative damage could result when the balance between the detoxification of the ROS products and the antioxidative system is altered (Kachout *et al.*, 2009). The tolerance of deleterious environmental stresses, such as heavy metals, is correlated with the increased capacity to scavenge or detoxify activated oxygen species (Kachout *et al.*, 2009). Boominathan and Doran (2003a,b) determined the role of antioxidative metabolism of heavy metal tolerance in *T. caerulescens*. They determined that superior antioxidant defenses, mainly catalase activity, may have an important role in the hyperaccumulator phenotype of *T. caerulescens*. Kachout et al (2010) determined the effects of Cu,

Ni, Pb and Zn on the antioxidative defense systems of *Atriplex* plants. They found that when the plants were exposed to different levels of metals, their dry matter production and shoot height decreased. Of the antioxidant enzymes, metal toxicity only diminished the levels of superoxide dismutase (SOD) and probably ascorbate peroxidase (APX) but increased the activity of catalase (CAT) and glutathione reductase (GR). The plants showed an intermediate level of tolerance to the metal stress conditions imposed. The antioxidative activity may be of fundamental significance for the*Atriplex* plants in their response against environmental stress.

Problems Associated with Plant Metal Tolerance

Soils enriched with metals are demanding on tolerant and accumulator plants. The costs associated with their adaptation to these sites are related to energy and resources allocations. When a metal tolerant or accumulator plant is growing in a metal contaminated soil, there is an increase in cost because the organism has to spend energy to counter the effects of the metals (Maestri *et al.,* 2010). Slow growth and low reproduction are the main characteristics of plants growing on metal enriched soils (Ernst, 2006; Ernst *et al.,* 2000). Haldane (1954) stated that costs are associated with the natural selection of new alleles. More energy and resources are required for the maintenance of the tolerance mechanisms at the cellular level. It has been demonstrated that tolerant plants have increased synthesis of complexing molecules in the cytosol. For example, metallothioneins and phytochelatins for the detoxification of metals such as As, Cd, and Cu. ATP are also needed for the active transport of metals across the plasma membrane and tonoplast. The synthesis of these agents withdraws N, S and energy from the primary metabolism (Ernst, 2006; Verkleij *et al.,* 1998). Energy is also required for the translocation of metals from root to shoot as well as for their allocation to various tissues and cell types. The reduced biomass of metal tolerant plants compared to their non metal tolerant ancestors might also be the result of less than favourable environmental conditions such as low water and nutrient supply. The diminished biomass and seed production might be the result of all costs associated with their survival to these metal contaminated sites, such as adaptation and environmental constraints (Ernst, 2006). Plants have an advantages growing on metal contaminated soil. As previously mentioned, there is a lack of competitive species on these sites. With high metal accumulation of metals in their aerial tissues, the "elemental hypothesis" speculates that hyperaccumulators can deter predators such as herbivores from feeding on them (Maestri *et al.*,2010; Vesk and Reichman 2009). However, some insects feed on hyperaccumulator plants and in turn accumulate the metals in their tissue which then aid in their defence against predators (Maestri *et al.*, 2010). This contradiction may explain why there is a mix of excluders, accumulator and hyperaccumulators growing on metal contaminated sites. Another advantage of hyperaccumulation is the elimination of competitive plants by further contaminating the surrounding soil by shedding their metal contaminated leaves (Maestri *et al.*, 2010).

Effects of Metals on Plant Population Diversity and Structure

Elevated accumulations of metals in soil and vegetation have been documented within short distances of the smelters compared to control sites (Nkongolo *et al.*, 2008; Gratton *et al.*, 2000). Several authors have reported differences in genetic structure of plants growing in contaminated areas (Vandeligt *et al.*, 2011; Nkongolo *et al.*, 2008; Scholz and Bergmann 1984). Enzymatic studies of Norway spruce (*Picea abies*) revealed genetic differences between groups of sensitive trees in polluted areas (Scholz and Bergmann 1984). It has been demonstrated that the evolution of heavy metal tolerant ecotypes occurs at an unexpectedly rapid rate (Wu *et al.*, 1975) and that despite founder effect and selection, in several cases, the recently established tolerant populations maintain a high level of variation and appear to be at least as variable as non tolerant populations. Observations of higher heterozygosity in tolerant plants of European beech (*Fagus sylvatica*) in Germany (Muller and Starck 1985), scots pine (*Pinus* sylvestris) in Germany and Great Britain (Geburek *et al.*, 1987), trembling aspen (*Populus tremuloides*) and red maple (*Acer rubrum*) in the United States (Berrang *et al.*, 1986) have been reported. Several studies, however, have reported the detection of bottleneck effects (Nordal *et al.*, 1999; Vekemans and Lefebre 1997; Mejnartowicz, 1983). Mejnatowicz

(1983) presented evidence of loss of genes and heterozygosity in tolerant Scots pines. The frequent lack of a bottleneck effect has been explained by different hypotheses: successive colonization events, a high number of tolerant plants in the primary populations, pollen flow from the neighboring populations, environmental heterogeneity and human disturbance (Nkongolo *et al.*, 2007).

Molecular analyses of several conifer and hardwood species clearly indicated that the exposure to metals for more than 30 years has no effect on genetic structure and diversity of early generations of*Picea mariana, P. glauca, Pinus banksiana, P. rubens, P. strobus,* and several hardwood populations in Northern Ontario (Narendrula *et al.*, 2012; Nkongolo *et al.*, 2012; Dobrzeniecka *et al.*, 2011; Vandeligt *et al.*, 2011). This lack of association between the level of genetic variation and metal content can be attributed to the long life span of these tree species. Table 1 shows similar level of genetic variabilities in pine populations growing in metal contaminated sites for more than 30 years compared to control in Northern Ontario, Canada. This is in contrast to data observed in herbaceous species such as *D. cespitosa* where a high level of metal accumulation reduced significantly the level of genetic variation (Table 2) (Nkongolo *et al.*, 2007). Metals impose severe stress on plants, especially in the rooting zone, which has led to the evolution of metal resistant ecotypes in several herbaceous species like *D. cespitosa* (Cox and Hutchinson 1980).

Table 1: Genetic variability parameters of *Pinus banksiana* populations growing in the Sudbury, Ontario (Canada) area based on ISSR data

Populations	P (%)	h	I	Ne	Na
Vale site 1 (metal contaminated)	31.25	0.1120	0.1653	1.2035	1.3125
Vale site 2 (metal contaminated	31.25	0.1171	0.1727	1.2061	1.3125
Xtrata 2 (metal contaminated)	27.08	0.0995	0.1467	1.1758	1.2708
Xtrata 3 (metal contaminated)	20.83	0.0630	0.0982	1.1004	1.2083
Vale Tailing (metal contaminated)	35.42	0.0977	0.1552	1.1514	1.3542

Temagami site (control)	29.17	0.0818	0.1284	1.1310	1.2917
Low Water Lake (control)	31.25	0.0812	0.1297	1.1256	1.3125
Mean	31.63	0.1001	0.1528	1.1679	1.3163

P represents percentage of polymorphic loci; h, Nei's gene diversity; I, Shannon's information index; Ne, effective number of alleles; Na, observed number of alleles.

Table 2: Genetic variability within *Deschampsia cespitosa* populations from Northern Ontario generated with ISSR primers

Region	Site	Polymorphism per site (%)	Mean polymorphism per region (%)
Sudbury	Coniston	72	74 (Sudbury)
(moderately contaminated)	Xtrata	92	
	Copper Cliff	67	
	Walden	65	
Cobalt	Cobalt-3	48	46 (Cobalt)
(highly contaminated)	Cobalt-4	46	
	Cobalt-5	44	
Manitoulin	Little Current	70	69 (Manitoulin)
(control)	Mississagi Lighthouse	68	

P represents percentage of polymorphic loci; Sudbury and Cobalt regions were moderately and highly contaminated with metals, respectively. Manitoulin Island region was not contaminated with metals and was used as a control region.

CONCLUSIONS

Plants play an essential role in the remediation of metal enriched soils. Coping mechanisms developed by some group of plants growing on metal contaminated soil facilitate the establishment of sustainable ecosystems in areas that would otherwise remain barren. A number of studies have been completed to explain the complex mechanisms involved in tolerance genotypes, and also the biological variability in their environmental adaptation. Depending on the circumstances, metal excluders or hyperaccumulators may be used to remediate polluted soil. Excluders may be useful for soil stabilization by preventing wind and water erosion and also by limiting the entry of heavy metals in the food chain. Metals can also be extracted by hyperaccumulators but since majority of these plants have low biomass, the extraction of metals from soil is very slow. The remediation of these sites using this technique may take up to hundreds of years. With genetic engineering, it may be possible to design the ideal plant prototype for the remediation of metal contamination in different environments. Many genes and mechanisms have been identified to have a role in tolerance and hyperaccumulation of metals. However, there is still a need for a better understanding of the mechanisms such as characterization of promoters of genes controlling metal tolerance and hyperaccumulation. This new knowledge would significantly contribute to a better understanding of the regulation and expression of different genes in hyperaccumulators. It is essential to mimic this regulation and expression of genes in high biomass non hyperaccumulators in order to obtain the hyperaccumulator phenotype.

ACKNOWLEDGEMENTS

We express our appreciation to the Natural Sciences and Engineering Research Council of Canada (NSERC) for a Postdoctoral Fellowship to M. Mehes-Smith.

REFERENCES

1. S. E Allen, 1989Chemical analysis of ecological materials. Blackwell Scientific Publications, Oxford. 368

2. Å. R Almås, C Kweyunga, and M. L. K Manoko, 2009Investigation of trace metal concentrations in soil, sediments and waters in the vicinity of "Geita Gold Mine" and "North Mara Gold Mine" in North West Tanzania (with Barrick Response). IPM-report.
3. J Arines, A Vilarino, and M Sainz, 1989Effect of Different Inocula of Vesicular-Arbuscular Mycorrhizal Fungi on Manganese Content and Concentration in Red-Clover (Trifolium pratense L) Plants. New Phytol. 112215219
4. A. G. L Assunçãno, W. M Bookum, H. J. M Nelissen, R Vooijs, H Schat, and W. H. O Ernst, 2003Differential metal-specific tolerance and accumulation patterns among Thlaspi caerulescens populations originating from different soil types. New Phytol. 159411419
5. A Assunçãno, B Pieper, J Vromans, P Lindhout, M Aarts, and H Schat, 2006Construction of a genetic linkage map of Thlaspi caerulescens and quantitative trait loci analysis of zinc accumulation. New Phytol. 1702132
6. A Assunçãno, P Martins, S De Folter, R Vooijs, H Schat, and M Aarts, 2001Elevated expression of metal transporter genes in three accessions of the metal hyperaccumulator Thlaspi caerulescens. Plant Cell and Environment 24217226
7. A. G. L Assunçãno, P Bleeker, ten Bookum, W.M., Vooijs, R. and Schat, H., 2008Intraspecific variation of metal preference patterns for hyperaccumulation in Thlaspi caerulescens: evidence from binary metal exposures. Plant Soil 303289299
8. K Axelsen, and M Palmgren, 2001Inventory of the superfamily of P-type ion pumps in Arabidopsis. Plant Physiol. 126696706
9. A. J. M Baker, and P. L Walker, 1990Ecophysiology of metal uptake by tolerant plants: Heavy metal uptake by tolerant plants. In: Shaw, A.J. (Editors). Evolutionary Aspects. CRC, Boca Raton.
10. A. J. M Baker, S. P Mcgrath, R. D Reeves, and J. A. C Smiith, 2000Metal hyperaccumulator plants: A review of the ecolgy and physiology of a biological resource for phytoremediation of metal-polluted soil. In: Terry, N. and Banuelos, G. (Editors). Phytoremediation of contaminated soil and water. CRC Press LLC, Boca Ratton. 85108
11. A. J. M Baker, 1987Metal tolerance. New Phytologist 106 (Suppl s1): 93-111.

12. A. F Bálint, G Kovacs, A Börner, G Galiba, and J Sutka, 2003Substitution analysis of seedling stage copper tolerance in wheat. Acta agronimica hungarica 51397404
13. A. F Bálint, M. S Röder, R Hell, G Galiba, and A Börner, 2007Mapping of QTLs affecting copper tolerance and the Cu, Fe, Mn and Zn contents in the shoots of wheat seedlings. Biol. Plant. 51129134
14. N Basic, N Salamin, C Keller, N Galland, and G Besnard, 2006Cadmium hyperaccumulation and genetic differentiation of Thlaspi caerulescens populations. Biochem. Syst. Ecol. 34667677
15. I Baxter, J Tchieu, M Sussman, M Boutry, M Palmgren, M Gribskov, J Harper, and K Axelsen, 2003Genomic comparison of P-type ATPase ion pumps in Arabidopsis and rice. Plant Physiol. 132618628
16. E. R Benitez, M Hajika, and R Takahashi, 2012Single-base substitution in 1Bgene is associated with a major QTL for seed cadmium concentration in soybean. J. Hered. 103: 278-286.
17. E. R Benitez, M Hajika, T Yamada, K Takahashi, N Oki, N Yamada, T Nakamura, and K Kanamaru, 2010A Major QTL Controlling Seed Cadmium Accumulation in Soybean. Crop Sci. 5017281734
18. P Berrang, D. F Karnosky, R. A Mickler, and J. P Bennett, 1986Natural selection for ozone tolerance in Populus tremuloides. Canadian Journal of Forest Research 1612141216
19. V Bert, I Bonnin, P Saumitou-laprade, P De Laguerie, and D Petit, 2002Do Arabidopsis halleri from nonmetallicolous populations accumulate zinc and cadmium more effectively than those from metallicolous populations? New Phytol. 1554757
20. V Bert, P Meerts, P Saumitou-laprade, P Salis, W Gruber, and N Verbruggen, 2003Genetic basis of Cd tolerance and hyperaccumulation in Arabidopsis halleri. Plant Soil 249918
21. G Besnard, N Basic, P Christin, D Savova-bianchi, and N Galland, 2009Thlaspi caerulescens (Brassicaceae) population genetics in western Switzerland: is the genetic structure affected by natural variation of soil heavy metal concentrations? New Phytol. 181974984
22. R Boominathan, and P Doran, 2003bCadmium tolerance and antioxidative defenses in hairy roots of the cadmium

hyperaccumulator, Thlaspi caerulescens. Biotechnol. Bioeng. 83158167

23. R Boominathan, and P Doran, 2003aOrganic acid complexation, heavy metal distribution and the effect of ATPase inhibition in hairy roots of hyperaccumulator plant species. J. Biotechnol. 101131146
24. A Boularbah, C Schwartz, G Bitton, W Aboudrar, A Ouhammou, and J. L Morel, 2006Heavy metal contamination from mining sites in South Morocco: 2. Assessment of metal accumulation and toxicity in plants. Chemosphere 63811817
25. A. D Bradshaw, 1991Genostasis and the Limits to Evolution. Philosophical Transactions of the Royal Society of London Series B-Biological Sciences 333289305
26. A. D Bradshaw, 1984Adaptation of plants to soils containing toxic metals- a test for conceit. Origins and development of adaptation, CIBA Foundation Symposium 102, Pitman, London, 419
27. M. G Bratteler, 2005Genetic architecture of traits associated with serpentine adaptation of Silene vulgaris (Caryophyllaceae). Eidgenoessische Technische Hochschule Zuerich (Switzerland), Switzerland. 91
28. C Broadhurst, R Chaney, J Angle, T Maugel, E Erbe, and C Murphy, 2004Simultaneous hyperaccumulation of nickel, manganese, and calcium in Alyssum leaf trichomes. Environ. Sci. Technol. 3857975802
29. A Brookes, J. C Collins, and D. A Thurman, 1981The mechanism of zinc tolerance in grasses. J. Plant Nutr. 3695705
30. R. R Brooks, 1998Plant that accumulate heavy metals. CAB International, Wallingford.
31. D. G Burke, K Watkins, and B. J Scott, 1990Manganese toxicity effects on visible symptoms, yield, manganese levels, and organic-acid levels in tolerant and sensitive wheat cultivars. Crop Sci. 30275280
32. I Cakmak, 2008Enrichment of cereal grains with zinc: Agronomic or genetic biofortification? Plant Soil 302117
33. A. K Chaturvedi, A Mishra, V Tiwari, and B Jha, 2012Cloning and transcript analysis of type 2 metallothionein gene (SbMT-2) from

extreme halophyte Salicornia brachiata and its heterologous expression in E. coli. Gene 499280287

34. D Ci, D Jiang, S Li, B Wollenweber, T Dai, and W Cao, 2012Identification of quantitative trait loci for cadmium tolerance and accumulation in wheat. Acta Physiol Plant 34191202
35. G Cieslinski, K. C. J Van Rees, A. M Szmigielska, G. S. R Krishnamurti, and P. M Huang, 1998Low-molecular-weight organic acids in rhizosphere soils of durum wheat and their effect on cadmium bioaccumulation. Plant Soil 203109117
36. J Clarke, D Leisle, and G Kopytko, 1997Inheritance of cadmium concentration in five durum wheat crosses. Crop Sci. 3717221726
37. S Clemens, 2006Toxic metal accumulation, responses to exposure and mechanisms of tolerance in plants. Biochimie 8817071719
38. C Cobbett, and P Goldsbrough, 2002Phytochelatins and metallothioneins: Roles in heavy metal detoxification and homeostasis. Annual Review of Plant Biology 53159182
39. I Colzi, S Doumett, Del Bubba, M., Fornaini, J., Arnetoli, M., Gabbrielli, R. and Gonnelli, C., 2011On the role of the cell wall in the phenomenon of copper tolerance in Silene paradoxa L. Environ. Exp. Bot. 727783
40. I Colzi, M Arnetoli, A Gallo, S Doumett, Del Bubba, M., Pignattelli, S., Gabbrielli, R. and Gonnelli, C., 2012Copper tolerance strategies involving the root cell wall pectins in Silene paradoxa L. Environ. Exp. Bot. 789198
41. M Courbot, G Willems, P Motte, S Arvidsson, N Roosens, P Saumitou-laprade, and N Verbruggen, 2007A Major Quantitative Trait Locus for Cadmium Tolerance in Arabidopsis halleri Colocalizes with HMA4, a Gene Encoding a Heavy Metal ATPase. Plant physiology 144: 1065.
42. R. M Cox, and T. C Hutchinson, 1980Multiple metal tolerances in the grass Deschampsia cespitosa (L.) Beauv. from the Sudbury smelting area. New Phytologist 84631647
43. P Coyle, J Philcox, L Carey, and A Rofe, 2002Metallothionein: The multipurpose protein. Cellular and Molecular Life Sciences 59627647
44. C Curie, G Cassin, D Couch, F Divol, K Higuchi, Le Jean, M., Misson, J., Schikora, A., Czernic, P. and Mari, S., 2009Metal

movement within the plant: contribution of nicotianamine and yellow stripe 1-like transporters. Annals of Botany 103111

45. A. X Deniau, B Pieper, Ten Bookum, W.M., Lindhout, P., Aarts, M.G.M. and Schat, H., 2006QTL analysis of cadmium and zinc accumulation in the heavy metal hyperaccumulator Thlaspi caerulescens. Theor. Appl. Genet. 113907920
46. Di BaccioD., Galla, G., Bracci, T., Andreucci, A., Barcaccia, G., Tognetti, R., Sebastiani, L. and Moshelion, M., 2011Transcriptome analyses of Populus × euramericana clone I-214 leaves exposed to excess zinc. Tree Physiol. 3112931308
47. DiDonatoR., Roberts, L., Sanderson, T., Eisley, R. and Walker, E., 2004Arabidopsis Yellow Stripe-Like2 (YSL2): a metal-regulated gene encoding a plasma membrane transporter of nicotianamine-metal complexes. Plant Journal 39403414
48. S Dobrzeniecka, K. K Nkongolo, P Michael, M Mehes-smith, and P Beckett, 2011Genetic Analysis of Black Spruce (Picea mariana) Populations from Dry and Wet Areas of a Metal-Contaminated Region in Ontario (Canada). Water Air and Soil Pollution 215117125
49. I Dufey, P Hakizimana, X Draye, S Lutts, and P Bertin, 2009QTL mapping for biomass and physiological parameters linked to resistance mechanisms to ferrous iron toxicity in rice. Euphytica 167143160
50. V Dushenkov, P Kumar, H Motto, and I Raskin, 1995Rhizofiltration-the use of Plants to Remove Heavy-Metals from Aqueous Streams. Environ. Sci. Technol. 2912391245
51. D. J Ecker, T. R Butt, and S. T Crooke, 1989Yeast Metallothionein-Gene-Function and Regulation by Metal-Ions. Met. Ions Biol. Syst. 25147169
52. J Entry, L Watrud, and M Reeves, 1999Accumulation of Cs-137 and Sr-90 from contaminated soil by three grass species inoculated with mycorrhizal fungi. Environmental Pollution 104449457
53. W. H. O Ernst, 2004Vegetation, organic matter and soil quality. In: Doelman, P. and Eijsackers, H.J.P. (Editors). Vital soil : Function, value, and properties. Amsterdam, Elsevier, 2004, 4198
54. W. H. O Ernst, 2006Evolution of metal tolerance in higher plants. Forest Snow and Landscape Research 80251274

55. W Ernst, H Nelissen, and Ten Bookum, W., 2000Combination toxicology of metal-enriched soils: physiological responses of a Zn- and Cd-resistant ecotype of Silene vulgaris on polymetallic soils. Environ. Exp. Bot. 435571
56. W. H. O Ernst, G Krauss, J. A. C Verkleij, and D Wesenberg, 2008Interaction of heavy metals with the sulphur metabolism in angiosperms from an ecological point of view. Plant Cell and Environment 31123143
57. F. W Fifield, and P. J Haines, 2000Environmental Analytical Chemistry. Wiley-Blackwell, 512
58. V Filatov, J Dowdle, N Smirnoff, B Ford-lloyd, H. J Newbury, and M Macnair, 2006Comparison of gene expression in segregating families identifies genes and genomic regions involved in a novel adaptation, zinc hyperaccumulation. Molecular Ecology 15: 3059.
59. V Filatov, J Dowdle, N Smirnoff, Ford Lloyd, B., Newbury, H.J. and Macnair, M.R., 2007A quantitative trait loci analysis of zinc hyperaccumulation in Arabidopsis halleri. New Phytologist 174: 590.
60. H Frérot, M Faucon, G Willems, C Godé, A Courseaux, A Darracq, N Verbruggen, and P Saumitou-laprade, 2010Genetic architecture of zinc hyperaccumulation in Arabidopsis halleri : the essential role of QTL x environment interactions. New Phytol. 187355367
61. U Galli, H Schuepp, and C Brunold, 1996Thiols in cadmium- and copper-treated maize (Zea mays L). Planta 198139143
62. G Ganeva, S Landjeva, and M Merakchijska, 2003Effects of chromosome substitutions on copper toxicity tolerance in wheat seedlings. Biol. Plant. 47621623
63. M Gaudet, F Pietrini, I Beritognolo, V Iori, M Zacchini, A Massacci, G. S Mugnozza, M Sabatti, and R Tognetti, 2011Intraspecific variation of physiological and molecular response to cadmium stress in Populus nigra L. Tree Physiol. 3113091318
64. T. H Geburek, F Scholz, W Knabe, and A Vornweg, 1987Genetic studies by isozyme gene loci on tolerance and sensitivity in an air polluted Pinus sylvestris field trial. Silvae Genetica 364953
65. D Gendre, P Czernic, G Conejero, K Pianelli, J Briat, M Lebrun, and S Mari, 2007TcYSL3, a member of the YSL gene family

from the hyper-accumulator Thlaspi caerulescens, encodes a nicotianamine-Ni/Fe transporter. Plant Journal 49115

66. D Gonzalez-mendoza, A. Q Moreno, and O Zapata-perez, 2007Coordinated responses of phytochelatin synthase and metallothionein genes in black mangrove, Avicennia germinans, exposed to cadmium and copper. Aquatic Toxicology 83306314
67. W Gratton, K Nkongolo, and G Spiers, 2000Heavy metal accumulation in soil and jack pine (Pinus banksiana) needles in Sudbury, Ontario, Canada. Bull. Environ. Contam. Toxicol. 64550557
68. A Gunes, G Soylemezoglu, A Inal, E. G Bagci, S Coban, and O Sahin, 2006Antioxidant and stomatal responses of grapevine (Vitis vinifera L.) to boron toxicity. Scientia Horticulturae 110279284
69. J. L Gustin, M. E Loureiro, D Kim, G Na, M Tikhonova, and D. E Salt, 2009MTP1-dependent Zn sequestration into shoot vacuole's suggests dual roles in Zn tolerance and accumulation in Zn hyperaccumulating plants. Plant Journal 5711161127
70. J. B. S Haldane, 1954The biochemistry of genetics. Allen & Unwin, London. 144
71. J Hall, 2002Cellular mechanisms for heavy metal detoxification and tolerance. J. Exp. Bot. 53111
72. M Hanikenne, and C Nouet, 2011Metal hyperaccumulation and hypertolerance: a model for plant evolutionary genomics. Curr. Opin. Plant Biol. 14252259
73. H Harmens, N. G. C. P. B Gusmao, P. R Denhartog, J. A. C Verkleij, and W. H. O Ernst, 1993Uptake and transport of zinc in zinc-sensitive and zinc-tolerant Silene vulgaris. J. Plant Physiol. 141309315
74. R. M Harrison, and M. B Chirgawi, 1989The assessment of air and soil as contributors of some trace-metals to vegetable plants.1. Use of a filtered air growth cabinet. Sci. Total Environ. 831334
75. Z Hassan, and M. G. M Aarts, 2011Opportunities and feasibilities for biotechnological improvement of Zn, Cd or Ni tolerance and accumulation in plants. Environ. Exp. Bot. 725363
76. V. H Hassinen, A. I Tervahauta, P Halimaa, M Plessl, S Peraniemi, H Schat, M. G. M Aarts, K Servomaa, and S. O Karenlampi,

2007Isolation of Zn-responsive genes from two accessions of the hyperaccumulator plant Thlaspi caerulescens. Planta 225977989

77. G Huang, and Y Wang, 2010Expression and characterization analysis of type 2 metallothionein from grey mangrove species (Avicennia marina) in response to metal stress. Aquatic Toxicology 998692
78. G Huang, and Y Wang, 2009Expression analysis of type 2 metallothioneingeneinmangrovespecies(Bruguieragymnorrhiza) under heavy metal stress. Chemosphere 7710261029
79. D Hussain, M Haydon, Y Wang, E Wong, S Sherson, J Young, J Camakaris, J Harper, and C Cobbett, 2004P-type ATPase heavy metal transporters with roles in essential zinc homeostasis in Arabidopsis. Plant Cell 1613271339
80. B. R Induri, D. R Ellis, G. T Slavov, T Yin, X Zhang, W Muchero, G. A Tuskan, and DiFazio, S.2012Identification of quantitative trait loci and candidate genes for cadmium tolerance in Populus. Tree Physiol. 32: 626-638.
81. S Ishikawa, N Ae, and M Yano, 2005Chromosomal regions with quantitative trait loci controlling cadmium concentration in brown rice (Oryza sativa). New Phytol. 168345350
82. S Ishikawa, T Abe, M Kuramata, M Yamaguchi, T Ando, T Yamamoto, and M Yano, 2010A major quantitative trait locus for increasing cadmium-specific concentration in rice grain is located on the short arm of chromosome 7. J. Exp. Bot. 61923934
83. S Jegadeesan, K Yu, V Poysa, E Gawalko, M. J Morrison, C Shi, and E Cober, 2010Mapping and validation of simple sequence repeat markers linked to a major gene controlling seed cadmium accumulation in soybean [Glycine max (L.) Merr]. Theor. Appl. Genet. 121:.
84. J Jeong, and M. L Guerinot, 2008Biofortified and bioavailable: The gold standard, for plant-based diets. Proc. Natl. Acad. Sci. U. S. A. 10517771778
85. S. S Kachout, Ben Mansoura, A., Leclerc, J.C., Mechergui, R., Rejeb, M.N. and Ouerghi, Z., 2009Effects of heavy metals on antioxidant activities of Atriplex hortensis and A. rosea. J. Food Agric. Environ. 7938945

86. M Kawachi, Y Kobae, H Mori, R Tomioka, Y Lee, and M Maeshima, 2009A Mutant Strain Arabidopsis thaliana that Lacks Vacuolar Membrane Zinc Transporter MTP1 Revealed the Latent Tolerance to Excessive Zinc. Plant and Cell Physiology 5011561170

87. S Kim, M Takahashi, K Higuchi, K Tsunoda, H Nakanishi, E Yoshimura, S Mori, and N Nishizawa, 2005Increased nicotianamine biosynthesis confers enhanced tolerance of high levels of metals, in particular nickel, to plants. Plant and Cell Physiology 4618091818

88. R. E Knox, C. J Pozniak, F. R Clarke, J. M Clarke, S Houshmand, and A. K Singh, 2009Chromosomal location of the cadmium uptake gene (Cdu1) in durum wheat. Genome 52741747

89. H Konno, T Nakato, S Nakashima, and K Katoh, 2005Lygodium japonicum fern accumulates copper in the cell wall pectin. J. Exp. Bot. 5619231931

90. H Konno, S Nakashima, and K Katoh, 2010Metal-tolerant moss Scopelophila cataractae accumulates copper in the cell wall pectin of the protonema. J. Plant Physiol. 167358364

91. Y Korshunova, D Eide, W Clark, M Guerinot, and H Pakrasi, 1999The IRT1 protein from Arabidopsis thaliana is a metal transporter with a broad substrate range. Plant Mol. Biol. 403744

92. U Krämer, 2005MTP1 mops up excess zinc in Arabidopsis cells. Trends Plant Sci. 10313315

93. U Krämer, J Cotter-howells, J Charnock, A Baker, and J Smith, 1996Free histidine as a metal chelator in plants that accumulate nickel. Nature 379635638

94. G Krishnamurti, G Cieslinski, P Huang, and K Vanrees, 1997Kinetics of cadmium release from soils as influenced by organic acids: Implication in cadmium availability. J. Environ. Qual. 26271277

95. N Lal, 2010Molecular mechanisms and genetic basis of heavy metal toxicity and tolerance in plants. In: Ashraf, M., Ozturk, M. and Ahmad, M.S.A. (Editors). Plant Adaptation and Phytoremediation. Springer Netherlands, 3558

96. J Lee, R. D Reeves, R. R Brooks, and T Jaffré, 1977Isolation and identification of a citrato-complex of nickel from nickel-accumulating plants. Phytochemistry 1615031505

97. J Levitt, 1980Responses of Plants to Environmental Stresses: Water, radiation, salt, and other stresses. Academic Press, New York. 607
98. T Li, X Yang, X Jin, Z He, P Stoffella, and Q Hu, 2005Root responses and metal accumulation in two contrasting ecotypes of Sedum alfredii Hance under lead and zinc toxic stress. Journal of Environmental Science and Health Part A-Toxic/hazardous Substances & Environmental Engineering 4010811096
99. H Ling, G Koch, H Baumlein, and M Ganal, 1999Map-based cloning of chloronerva, a gene involved in iron uptake of higher plants encoding nicotianamine synthase. Proc. Natl. Acad. Sci. U. S. A. 9670987103
100. C. P Liu, Z. G Shen, and X. D Li, 2007Accumulation and detoxification of cadmium in Brassica pekinensis and B. chinensis. Biol. Plant. 51116120
101. M. R Macnair, V Bert, S. B Huitson, P Saumitou-laprade, and D Petit, 1999Zinc tolerance and hyperaccumulation are genetically independent characters. Proceedings of the Royal Society of London. Series B, Biological Sciences, 26621752179
102. M. R Macnair, 1993The genetics of metal tolerance in vascular plants. New Phytologist 124541559
103. M. R Macnair, 1987Heavy metal tolerance in plants: a model evolutionary system. Trends in Ecology and Evolution 2354359
104. E Maestri, and M Marmiroli, 2012Genetic and molecular aspects of metal tolerance and hyperaccumulation. In: Gupta, D.K. and Sandalio, L.M. (Editors). Metal toxicity in plants: Perception, signaling and remediation. Sprigner-Verlag Berlin Heidelberg, 4163
105. E Maestri, M Marmiroli, G Visioli, and N Marmiroli, 2010Metal tolerance and hyperaccumulation: Costs and trade-offs between traits and environment. Environ. Exp. Bot. 68113
106. T Maitani, H Kubota, K Sato, and T Yamada, 1996The composition of metals bound to class III metallothionein (phytochelatin and its desglycyl peptide) induced by various metals in root cultures of Rubia tinctorum. Plant Physiol. 11011451150
107. M Marmiroli, F Pietrini, E Maestri, M Zacchini, N Marmiroli, A Massacci, and R Tognetti, 2011Growth, physiological

and molecular traits in Salicaceae trees investigated for phytoremediation of heavy metals and organics. Tree Physiol. 3113191334

108. H Marschner, 1995Mineral nutrition of higher plants. Edited by Horst Marschner. second edition. London; San Diego, Academic Press. 899
109. M Martínez, P Bernal, C Almela, D Velez, P Garcia-agustin, R Serrano, and J Navarro-avino, 2006An engineered plant that accumulates higher levels of heavy metals than Thlaspi caerulescens, with yields of 100 times more biomass in mine soils. Chemosphere 64478485
110. E Martinoia, M Klein, M Geisler, L Bovet, C Forestier, U Kolukisaoglu, B Muller-rober, and B Schulz, 2002Multifunctionality of plant ABC transporters- more than just detoxifiers. Planta 214345355
111. J. E Mayer, W. H Pfeiffer, and P Beyer, 2008Biofortified crops to alleviate micronutrient malnutrition. Curr. Opin. Plant Biol. 11166170
112. N. M Mayowa, and T. E Miller, 1991The Genetics of Tolerance to High Mineral Concentrations in the Tribe Triticeae- a Review and Update. Euphytica 57175185
113. L Mejnartowicz, 1983Changes in genetic structure of Scotch pine (Pinus sylvestris L.) population affected by industrial emission of fluoride and sulphur dioxide. Genetica Polonica 244150
114. N Mganga, M. L. K Manoko, and Z. K Rulangaranga, 2011Classification of plants according to their heavy metal content around north mara gold mine, Tanzania: Implication for phytoremediation. Tanzania Journal of Science 37109119
115. R Millaleo, M Reyes- Diaz, A. G Ivanov, M. L Mora, and M Alberdi, 2010Manganese as essential and toxic element for plants: Transport, accumulation and resistance mechanisms. Journal of Soil Science and Plant Nutrition 10470481
116. R Mills, G Krijger, P Baccarini, J Hall, and L Williams, 2003Functional expression of AtHMA4, a 1BATPase of the Zn/Co/Cd/Pb subclass. Plant Journal 35: 164-176.
117. M. J Milner, E Craft, N Yamaji, E Koyama, J. F Ma, and L. V Kochian, 2012Characterization of the high affinity Zn transporter from Noccaea caerulescens, NcZNT1, and dissection of its promoter

for its role in Zn uptake and hyperaccumulation. New Phytol. 195113123

118. B Montanini, D Blaudez, S Jeandroz, D Sanders, and M Chalot, 2007Phylogenetic and functional analysis of the Cation Diffusion Facilitator (CDF) family: improved signature and prediction of substrate specificity. BMC Genomics 8: 107.
119. R Narendrula, K. K Nkongolo, P Beckett, and G Spiers, 2012Total and bioavailable metals in two contrasting mining regions (Sudbury in Canada and Lubumbashi in DR-Congo): relation to genetic variation in plant populations. Chemistry and Ecology In press.
120. K Neumann, DrogeLaser, W., Kohne, S. and Broer, I., 1997Heat treatment results in a loss of transgene-encoded activities in several tobacco lines. Plant Physiol. 115939947
121. H Nian, Z Yang, S Ahn, Z Cheng, and H Matsumoto, 2002A comparative study on the aluminium- and copper-induced organic acid exudation from wheat roots. Physiol. Plantarum 116328335
122. S Nishida, C Tsuzuki, A Kato, A Aisu, J Yoshida, and T Mizuno, 2011AtIRT1, the primary iron uptake transporter in the root, mediates excess nickel accumulation in Arabidopsis thaliana. Plant and Cell Physiology 5214331442
123. K. K Nkongolo, M Mehes, A Deck, and P Michael, 2007Metal content in soil and genetic variation in Deschampsia cespitosa populations from Northern Ontario (Canada): application of ISSR markers. European Journal of Genetics Toxicology March: 138
124. K. K Nkongolo, R Narendrula, M Mehes-smith, S Dobrzeniecka, K Vandeligt, M Ranger, and P Beckett, 2012Genetic Sustainability of fragmented conifer populations from stressed areas in Northern Ontario (Canada): Application of molecular markers. In: Blanco, J.A. and Lo, Y.L. (Editors). Forest Ecosystems- More than just trees. Intech publisher, 315336
125. K. K Nkongolo, A Vaillancourt, S Dobrzeniecka, M Mehes, and P Beckett, 2008Metal content in soil and black spruce (Picea mariana) trees in the Sudbury region (Ontario, Canada): Low concentration of arsenic, cadmium, and nickel detected near smelter sources. Bull. Environ. Contam. Toxicol. 80107111

126. I Nordal, K. B Haraldsen, A Ergon, and A. B Eriksen, 1999Copper resistance and genetic diversity in Lychnis alpina (Caryophyllaceae) populations on mining sites. Folia Geobotanica 34471481

127. R Pal, and J. P. N Rai, 2010Phytochelatins: peptides involved in heavy metal detoxification. Appl. Biochem. Biotechnol. 160945963

128. R Palmiter, 1998The elusive function of metallothioneins. Proc. Natl. Acad. Sci. U. S. A. 9584288430

129. N Pandey, and C Sharma, 2002Effect of heavy metals Co2+, Ni2+ and Cd2+ on growth and metabolism of cabbage. Plant Science 163753758

130. A Papoyan, and L Kochian, 2004Identification of Thlaspi caerulescens genes that may be involved in heavy metal hyperaccumulation and tolerance. Characterization of a novel heavy metal transporting ATPase. Plant Physiol. 13638143823

131. J Pelloux, C Rusterucci, and E. J Mellerowicz, 2007New insights into pectin methylesterase structure and function. Trends Plant Sci. 12267277

132. N Pence, P Larsen, S Ebbs, D Letham, M Lasat, D Garvin, D Eide, and L Kochian, 2000The molecular physiology of heavy metal transport in the Zn/Cd hyperaccumulator Thlaspi caerulescens. Proc. Natl. Acad. Sci. U. S. A. 9749564960

133. L Perfus-barbeoch, N Leonhardt, A Vavasseur, and C Forestier, 2002Heavy metal toxicity: cadmium permeates through calcium channels and disturbs the plant water status. Plant Journal 32539548

134. K Pianelli, S Mari, L Marques, M Lebrun, and P Czernic, 2005Nicotianamine over-accumulation confers resistance to nickel in Arabidopsis thaliana. Transgenic Res. 14739748

135. A Pich, R Manteuffel, S Hillmer, G Scholz, and W Schmidt, 2001Fe homeostasis in plant cells: Does nicotianamine play multiple roles in the regulation of cytoplasmic Fe concentration? Planta 213967976

136. J Pintro, J Barloy, and P Fallavier, 1997Effects of low aluminum activity in nutrient solutions on the organic acid concentrations in maize plants. J. Plant Nutr. 20601611

137. I. D Pulford, and C Watson, 2003Phytoremediation of heavy metal-contaminated land by trees- a review. Environ. Int. 29529540
138. K. H Richau, and H Schat, 2009Intraspecific variation of nickel and zinc accumulation and tolerance in the hyperaccumulator Thlaspi caerulescens. Plant Soil 314253262
139. P Rotkittikhun, M Kruatrachue, R Chaiyarat, C Ngernsansaruay, P Pokethitiyook, A Paijitprapaporn, and A. J. M Baker, 2006Uptake and accumulation of lead by plants from the Bo Ngam lead mine area in Thailand. Environmental Pollution 144681688
140. M. L Salin, 1988Toxic oxygen species and protective systems of the chloroplast. Physiol. Plantarum 72681689
141. D. E Salt, and W. E Rauser, 1995Mgatp-dependent transport of phytochelatins across the tonoplast of oat roots. Plant Physiol. 10712931301
142. D. E Salt, R. D Smith, and I Raskin, 1998Phytoremediation. Annu. Rev. Plant Physiol. Plant Mol. Biol. 49643668
143. D. E Salt, R. C Prince, I. J Pickering, and I Raskin, 1995Mechanisms of cadmium mobility and accumulation in indian mustard. Plant Physiol. 10914271433
144. D Salt, R Prince, A Baker, I Raskin, and I Pickering, 1999Zinc ligands in the metal hyperaccumulator Thlaspi caerulescens as determined using X-ray absorption spectroscopy. Environ. Sci. Technol. 33713717
145. R Sanchez-fernandez, T Davies, J Coleman, and P Rea, 2001The Arabidopsis thaliana ABC protein superfamily, a complete inventory. J. Biol. Chem. 2763023130244
146. G Schaaf, A Honsbein, A. R Meda, S Kirchner, D Wipf, and N Von Wiren, 2006AtIREG2 encodes a tonoplast transport protein involved in iron-dependent nickel detoxification in Arabidopsis thaliana roots. J. Biol. Chem. 2812553225540
147. H Schat, L Mercè, and R Bernhard, 2000Metal-specific patterns of tolerance, uptake and transport of heavy metals in hyperaccumulating and nonhyperaccumulating metallophytes. In: Terry, N. and Banuelos, G. (Editors). Phytoremediation of Contaminated Soil and Water. CRC Press, Boca Raton. 171188
148. H Schat, E Kuiper, W Tenbookum, and R Vooijs, 1993A General-model, for the genetic-control of copper tolerance in Silene

Vulgaris- evidence from crosses between plants from different tolerant populations. Heredity 70142147

149. H Schat, R Vooijs, and E Kuiper, 1996Identical major gene loci for heavy metal tolerances that have independently evolved in different local populations and subspecies of Silene vulgaris. Evolution 5018881895
150. I Schmidke, and U. W Stephan, 1995Transport of metal micronutrients in the phloem of castor bean (Ricinus communis) seedlings. Physiol. Plantarum 95147153
151. F Scholz, and F Bergmann, 1984Selection pressure by air-pollution as studied by isozyme-gene-systems in norway spruce exposed to sulfur-dioxide. Silvae Genet. 33238241
152. K Sekhar, B Priyanka, V. D Reddy, and K. V Rao, 2011Metallothionein 1 (CcMT1) of pigeonpea (Cajanus cajan L.) confers enhanced tolerance to copper and cadmium in Escherichia coli and Arabidopsis thaliana. Environ. Exp. Bot. 72131139
153. S Shojima, N. K Nishizawa, S Fushiya, S Nozoe, T Irifune, and S Moris, 1990Biosynthesis of phytosiderophores- invitro biosynthesis of 2′-deoxymugineic acid from L-methionine and nicotianamine. Plant Physiol. 9314971503
154. S Smith, and M Macnair, 1998Hypostatic modifiers cause variation in degree of copper tolerance in Mimulus guttatus. Heredity 80760768
155. R Solanki, and R Dhankhar, 2011Biochemical changes and adaptive strategies of plants under heavy metal stress. Biologia 66195204
156. U. W Stephan, I Schmidke, and A Pich, 1994Phloem translocation of Fe, Cu, Mn, and Zn in ricinus seedlings in relation to the concentrations of nicotianamine, an endogenous chelator of divalent metal-ions, in different seedling parts. Plant Soil 165181188
157. E. R Still, and R. J. P Williams, 1980Potential methods for selective accumulation of nickel (II) ions by plants. J. Inorg. Biochem. 133540
158. M Takahashi, Y Terada, I Nakai, H Nakanishi, E Yoshimura, S Mori, and N Nishizawa, 2003Role of nicotianamine in the intracellular delivery of metals and plant reproductive development. Plant Cell 1512631280

159. P Tanhuanpää, R Kalendar, A. H Schulman, and E Kiviharju, 2007A major gene for grain cadmium accumulation in oat (Avena sativa L.). Genome 50588594
160. K Tezuka, H Miyadate, K Katou, I Kodama, S Matsumoto, T Kawamoto, S Masaki, H Satoh, M Yamaguchi, K Sakurai, H Takahashi, N Satoh-nagasawa, A Watanabe, T Fujimura, and H Akagi, 2010A single recessive gene controls cadmium translocation in the cadmium hyperaccumulating rice cultivar Cho-Ko-Koku. Theor. Appl. Genet. 12011751182
161. Y Tong, R Kneer, and Y Zhu, 2004Vacuolar compartmentalization: a second-generation approach to engineering plants for phytoremediation. Trends Plant Sci. 979
162. D Ueno, E Koyama, I Kono, T Ando, M Yano, and J. F Ma, 2009Identification of a novel major quantitative trait locus controlling distribution of cd between roots and shoots in rice. Plant and Cell Physiology 5022232233
163. van de MortelJ.E., Almar Villanueva, L., Schat, H., Kwekkeboom, J., Coughlan, S., Moerland, P.D., van Themaat, E.V.L., Koornneef, M. and Aarts, M.G.M., 2006Large expression differences in genes for iron and zinc homeostasis, stress response, and lignin biosynthesis distinguish roots of Arabidopsis thaliana and the related metal hyperaccumulator Thlaspi caerulescens. Plant Physiol. 14211271147
164. N. A Van Hoof, V. H Hassinen, H. W Hakvoort, K. F Ballintijn, H Schat, J. A Verkleij, W. H Ernst, S. O Karenlampi, and A. I Tervahauta, 2001Enhanced copper tolerance in Silene vulgaris (Moench) Garcke populations from copper mines is associated with increased transcript levels of a 2b-type metallothionein gene. Plant Physiol. 12615191526
165. K. K Vandeligt, K. K Nkongolo, M Mehes, and P Beckett, 2011Genetic analysis of Pinus banksiana and Pinus resinosa populations from stressed sites contaminated with metals in Northern Ontario (Canada). Chem. Ecol. 27369380
166. X Vekemans, and C Lefebre, 1997The evolution of heavy metal tolerant populations in Armenia maritime: evidence from allozyme variation and reproductive barriers. Journal of Evolutionary Biology 10175191

167. N Verbruggen, C Hermans, and H Schat, 2009Molecular mechanisms of metal hyperaccumulation in plants. New Phytol. 182781781

168. J Verkleij, P Koevoets, M Blake-kalff, and A Chardonnens, 1998Evidence for an important role of the tonoplast in the mechanism of naturally selected zinc tolerance in Silene vulgaris. J. Plant Physiol. 153188191

169. P Vesk, and S Reichman, 2009Hyperaccumulators and herbivores-a bayesian meta-analysis of feeding choice trials. J. Chem. Ecol. 35289296

170. G Visioli, S Vincenzi, M Marmiroli, and N Marmiroli, 2012Correlation between phenotype and proteome in the Ni hyperaccumulator Noccaea caerulescens subsp. caerulescens. Environmental and Experimental Botany 77156164

171. G Visioli, A Pirondini, A Malcevschi, and N Marmiroli, 2010Comparison of protein variations in Thlaspi Caerulescens populations from metalliferous and non-metalliferous soils. Int. J. Phytoremediation 12805819

172. R Vogelilange, and G. J Wagner, 1990Subcellular-localization of cadmium and cadmium-binding peptides in tobacco-leaves-implication of a transport function for cadmium-binding peptides. Plant Physiol. 9210861093

173. S Wei, Q Zhou, and X Wang, 2005Identification of weed plants excluding the uptake of heavy metals. Environ. Int. 31829834

174. W Wenzel, M Bunkowski, M Puschenreiter, and O Horak, 2003Rhizosphere characteristics of indigenously growing nickel hyperaccumulator and excluder plants on serpentine soil. Environmental Pollution 123131138

175. G Willems, D. B Dräger, M Courbot, C Godé, N Verbruggen, and P Saumitou-laprade, 2007The Genetic Basis of Zinc Tolerance in the metallophyte Arabidopsis halleri ssp. halleri (Brassicaceae): An Analysis of Quantitative Trait Loci. Genetics 176659674

176. H Wong, T Sakamoto, T Kawasaki, K Umemura, and K Shimamoto, 2004Down-regulation of metallothionein, a reactive oxygen scavenger, by the small GTPase OsRac1 in rice. Plant Physiol. 13514471456

177. L Wu, A. D Bradshaw, and D. A Thurman, 1975The potential for evolution of heavy metal tolerance in plants. Heredity 34165187

178. H. L Xie, R. F Jiang, F. S Zhang, S. P Mcgrath, and F. J Zhao, 2009Effect of nitrogen form on the rhizosphere dynamics and uptake of cadmium and zinc by the hyperaccumulator Thlaspi caerulescens. Plant Soil 318205215
179. J Xiong, Z He, D Liu, Q Mahmood, and X Yang, 2008The role of bacteria in the heavy metals removal and growth of Sedum alfredii Hance in an aqueous medium. Chemosphere 70489494
180. L Xu, L Wang, Y Gong, W Dai, Y Wang, X Zhu, T Wen, and L Liu, 2012Genetic linkage map construction and QTL mapping of cadmium accumulation in radish (Raphanus sativus L.). Theoretical and Applied Genetics 125659670
181. X Yang, Y Feng, Z He, and P Stoffella, 2005bMolecular mechanisms of heavy metal hyperaccumulation and phytoremediation. Journal of Trace Elements in Medicine and Biology 18339353
182. X Yang, X Jin, Y Feng, and E Islam, 2005aMolecular mechanisms and genetic basis of heavy metal tolerance/hyperaccumulation in plants. Journal of Integrative Plant Biology 4710251035
183. C Yong, and L. Q Ma, 2002Metal tolerance, accumulation, and detoxification in plants with emphasis on arsenic in terrestrial plants. In: Biogeochemistry of Environmentally Important Trace Elements. American Chemical Society, 95114
184. M Zacchini, F Pietrini, Scarascia Mugnozza, G., Iori, V., Pietrosanti, L. and Massacci, A., 2009Metal tolerance, accumulation and translocation in poplar and willow clones treated with cadmium in hydroponics. Water Air Soil Pollut. 1972334
185. M. H Zenk, 1996Heavy metal detoxification in higher plants- a review. Gene 1792130
186. H Zha, R Jiang, F Zhao, R Vooijs, H Schat, J Barker, and S Mcgrath, 2004Co-segregation analysis of cadmium and zinc accumulation in Thlaspi caerulescens interecotypic crosses. New Phytol. 163299312
187. Z. S Zhou, H. Q Zeng, Z. P Liu, and Z. M Yang, 2012Genome-wide identification of Medicago truncatula microRNAs and their targets reveals their differential regulation by heavy metal. Plant Cell and Environment 358699

Chapter 3

A Review on Heavy Metals (As, Pb, And Hg) Uptake by Plants through Phytoremediation

Bieby Voijant Tangahu[1], Siti Rozaimah Sheikh Abdullah[2], Hassan Basri[1], Mushrifah Idris[3], Nurina Anuar[2], and Muhammad Mukhlisin[1]

[1]Department of Civil and Structural Engineering, Faculty of Engineering and Built Environment, Universiti Kebangsaan Malaysia, 43600 Bangin, Malaysia

[2]Department of Chemical Engineering, Faculty of Engineering and Built Environment, Universiti Kebangsaan Malaysia, 43600 Bangin, Malaysia

[3]Tasik Chini Reasearch Centre, Faculty of Science and Technology, Universiti Kebangsaan Malaysia, 43600 Bangin, Malaysia

ABSTRACT

Heavy metals are among the most important sorts of contaminant in the environment. Several methods already used to clean up the environment from these kinds of contaminants, but most of them are costly and difficult to get optimum results. Currently, phytoremediation is an effective and affordable technological solution used to extract or remove inactive metals and metal pollutants from contaminated soil and water. This technology is environmental friendly and potentially cost effective. This paper aims to compile some information about heavy metals of arsenic, lead, and mercury (As, Pb, and Hg) sources, effects and their treatment. It also reviews deeply about phytoremediation technology, including the heavy metal uptake mechanisms and several research studies associated about the topics. Additionally, it describes several sources and the effects of As, Pb, and Hg on the environment, the advantages of this kind of technology for reducing them, and also heavy metal uptake mechanisms in phytoremediation technology as well as the factors affecting the uptake mechanisms. Some recommended plants which are commonly used in phytoremediation and their capability to reduce the contaminant are also reported.

INTRODUCTION

Heavy metals are among the contaminants in the environment. Beside the natural activities, almost all human activities also have potential contribution to produce heavy metals as side effects. Migration of these contaminants into noncontaminated areas as dust or leachates through the soil and spreading of heavy metals containing sewage sludge are a few examples of events contributing towards contamination of the ecosystems [1].

Several methods are already being used to clean up the environment from these kinds of contaminants, but most of them are costly and far away from their optimum performance. The chemical technologies generate large volumetric sludge and increase the costs [2]; chemical and thermal methods are both technically difficult and expensive that all of these methods can also degrade the valuable component of soils [3]. Conventionally, remediation of heavy-metal-contaminated soils involves either onsite management or excavation and subsequent disposal to a landfill site. This method of disposal solely shifts the contamination problem elsewhere along with the hazards associated with transportation of contaminated soil and migration of contaminants from landfill into an adjacent environment. Soil washing for removing contaminated soil is an alternative way to excavation and disposal to landfill. This method is very costy and produces a residue rich in heavy metals, which will require further treatment. Moreover, these physio-chemical technologies used for soil remediation render the land usage as a medium for plant growth, as they remove all biological activities [1].

Recent concerns regarding the environmental contamination have initiated the development of appropriate technologies to assess the presence and mobility of metals in soil [4], water, and wastewater. Presently, phytoremediation has become an effective and affordable technological solution used to extract or remove inactive metals and metal pollutants from contaminated soil. Phytoremediation is the use of plants to clean up a contamination from soils, sediments, and water. This technology is environmental friendly and potentially costeffective. Plants with exceptional metal-accumulating capacity are known as hyperaccumulator plants [5]. Phytoremediation takes the advantage of the unique and selective uptake capabilities of plant root systems, together with the translocation, bioaccumulation, and contaminant degradation abilities of the entire plant body [3].

Many species of plants have been successful in absorbing contaminants such as lead, cadmium, chromium, arsenic, and various radionuclides from soils. One of phytoremediation categories, phytoextraction, can be used to remove heavy metals from soil using its ability to uptake metals which are essential for plant growth (Fe, Mn, Zn, Cu, Mg, Mo, and Ni). Some metals with unknown biological function (Cd, Cr, Pb, Co, Ag, Se, Hg) can also be accumulated [5].

The objectives of this paper are to discuss the potential of phytoremediation technique on treating heavy metal-contaminated side, to provide a brief view about heavy metals uptake mechanisms by plant, to give some description about the performance of several types of plants to uptake heavy metals and to describe about the fate of heavy metals in plant tissue, especially on arsenic (As), lead (Pb), and mercury (Hg). This study is related to a research project that aims to identify potential plants in tropical country such as Malaysia which can uptake heavy metal contaminants from petrochemical wastewater.

HEAVY METALS: SOURCES AND EFFECT IN THE ENVIRONMENT

Heavy metals are conventionally defined as elements with metallic properties and an atomic number >20. The most common heavy metal contaminants are Cd, Cr, Cu, Hg, Pb, and Zn. Metals are natural components in soil [6]. Some of these metals are micronutrients necessary for plant growth, such as Zn, Cu, Mn, Ni, and Co, while others have unknown biological function, such as Cd, Pb, and Hg [1].

Metal pollution has harmful effect on biological systems and does not undergo biodegradation. Toxic heavy metals such as Pb, Co, Cd can be differentiated from other pollutants, since they cannot be biodegraded but can be accumulated in living organisms, thus causing various diseases and disorders even in relatively lower concentrations [7]. Heavy metals, with soil residence times of thousands of years, pose numerous health dangers to higher organisms. They are also known to have effect on plant growth, ground cover and have a negative impact on soil microflora [8]. It is well known that heavy metals cannot be chemically degraded and need to be physically removed or be transformed into nontoxic compounds [1].

Arsenic (As)

Arsenic (atomic number 33) is a silver-grey brittle crystalline solid with atomic weight of 74.9, specific gravity 5.73, melting point 817°C (at 28 atm), boiling point 613°C, and vapor pressure 1 mm Hg at 372°C [9]. Arsenic is a semimetallic element with the chemical symbol "As". Arsenic is odorless and tasteless. Arsenic can combine with other elements to form inorganic and organic arsenicals [10]. In the environment, arsenic is combined with oxygen, chlorine, and sulfur to form inorganic arsenic compounds. Inorganic arsenic compounds are mainly used to preserve wood. Organic arsenic compounds are used as pesticides, primarily on cotton plants [11].

Arsenic exists in the −3, 0, +3, and +5 valence oxidation states [9], and in a variety of chemical forms in natural waters and sediments [12]. Environmental forms include arsenious acids (H_3AsO_3, H_3AsO_3, $H_3AsO_3^{2-}$), arsenic acids (H_3AsO_4, $H_3AsO_4^-$, $H_3AsO_4^{2-}$), arsenites, arsenates, methylarsenic acid, dimethylarsinic acid, and arsine. Two most common forms in natural waters arsenite (AsO_3^{3-}) and inorganic arsenate (AsO_4^{3-}), referred as As^{3+} and As^{5+} [9]. From both the biological and the toxicological points of view, arsenic compounds can be classified into three major groups. These groups are inorganic arsenic compounds, organic arsenic compounds, and arsine gas [13].

It is a hard acid and preferentially complexes with oxides and nitrogen. Trivalent arsenites predominate in moderately reducing anaerobic environments such as groundwater [9]. The most common trivalent inorganic arsenic compounds are arsenic trioxide, sodium arsenite, and arsenic trichloride [13]. Trivalent (+3) arsenates include $As(OH)_3$, $As(OH)_4^-$, AsO_2OH^{2-}, and AsO_3^{3-} [9]. Arsenite ($As(OH)_3$, As^{3+}) is predominant in reduced redox potential conditions [12].

Arsenic is one of the contaminants found in the environment which is notoriously toxic to man and other living organisms [14]. It is a highly toxic element that exists in various species, and the toxicity of arsenic depends on its species. The pH, redox conditions, surrounding mineral composition, and microbial activities affect the form (inorganic or organic) and the oxidation state of arsenic. It is generally accepted that the inorganic species, arsenite [As^{3+}] and arsenate [As^{5+}], are the predominant species in most environments, although the organic ones might also be present [15].

In general, inorganic compounds of arsenic are regarded as more highly toxic than most organic forms which are less toxic [10, 14, 16, 17]. The trivalent compounds (arsenites) are more toxic than the pentavalent compounds (arsenates) [16, 17]. It has been reported that As^{3+} is 4 to 10 times more soluble in water than As^{5+}. However, the trivalent methylated arsenic species have been found to be more toxic than inorganic arsenic because they are more efficient at causing DNA breakdown [17]. Although As^{5+} tends to be less toxic compared to of As^{3+}, it is thermodynamically more stable due to it predominates under normal conditions and becomes the cause of major contaminant in ground water [14]. Arsenate which is in the pentavalent state (As^{5+}) is also considered to be toxic and carcinogenic to human [18].

Lead (Pb)

Lead (Pb), with atomic number 82, atomic weight 207.19, and a specific gravity of 11.34, is a bluish or silvery-grey metal with a melting point of 327.5°C and a boiling point at atmospheric pressure of 1740°C. It has four naturally occurring isotopes with atomic weights 208, 206, 207 and 204 (in decreasing order of abundance). Despite the fact that lead has four electrons on its valence shell, its typical oxidation state is +2 rather than +4, since only two of the four electrons ionize easily. Apart from nitrate, chlorate, and chloride, most of the inorganic salts of $lead^{2+}$ have poor solubility in water [19]. Lead (Pb) exists in many forms in the natural sources throughout the world and is now one of the most widely and evenly distributed trace metals. Soil and plants can be contaminated by lead from car exhaust, dust, and gases from various industrial sources.

Pb^{2+} was found to be acute toxic to human beings when present in high amounts. Since Pb^{2+} is not biodegradable, once soil has become contaminated, it remains a long-term source of Pb^{2+} exposure. Metal pollution has a harmful effect on biological systems and does not undergo biodegradation [7].

Soil can be contaminated with Pb from several other sources such as industrial sites, from leaded fuels, old lead plumbing pipes, or even old orchard sites in production where lead arsenate is used. Lead accumulates in the upper 8 inches of the soil and is highly immobile. Contamination is long-term. Without remedial action, high soil lead levels will never return to normal [20].

In the environment, lead is known to be toxic to plants, animals, and microorganisms. Effects are generally limited to especially contaminated areas [21]. Pb contamination in the environment exists as an insoluble form, and the toxic metals pose serious human health problem, namely, brain damage and retardation [5].

Mercury (Hg)

Mercury is a naturally occurring metal that is present in several forms. Metallic mercury is shiny, silver-white, odorless liquid. Mercury combines with other elements, such as chlorine, sulfur, or oxygen, to form inorganic mercury compounds or salts, which are usually white powders or crystals. Mercury also combines with carbon to make organic mercury compounds [22]. Mercury, which has the lowest melting point (−39°C) of all the pure metals, is the only pure metal that is liquid at room temperature. However, due to its several physical and chemical advantages such as its low boiling point (357°C) and easy vaporization, mercury is still an important material in many industrial products [23]. As any other metal, mercury could occur in the soil in various forms. It dissolves as free ion or soluble complex and is nonspecifically adsorbed by binding mainly due to the electrostatic forces, chelated, and precipitated as sulphide, carbonate, hydroxide, and phosphate. There are three soluble forms of Hg in the soil environment. The most reduced is Hg^0 metal with the other two forms being ionic of mercurous ion Hg_2^{2+} and mercuric ion Hg^{2+}, in oxidizing conditions especially at low pH. Hg^+ion is not stable under environmental conditions since it dismutates into Hg^0 and Hg^{2+}. A second potential route for the conversion of mercury in the soil is methylation to methyl or dimethyl mercury by anaerobic bacteria [24].

Mercury is a persistent environmental pollutant with bioaccumulation ability in fish, animals, and human beings [23]. Mercury salts and organomercury compounds are among the most poisonous substances in our environment. The mechanism and extent of toxicity depend strongly on the type of compound and the redox state of mercury [25].

Environmental contamination due to mercury is caused by several industries, petrochemicals, minings, painting, and also by agricultural sources such as fertilizer and fungicidal sprays [26]. Some of the more common sources of mercury found throughout the environment include but may not be limited to the household bleach, acid, and caustic chemicals (e.g., battery acid, household lye, muriatic acid (hydrochloric acid), sodium hydroxide, and sulfuric acid), instrumentation containing mercury (e.g., medical instruments, thermometers, barometers, and manometers), dental amalgam (fillings), latex paint (manufactured prior to 1990), batteries, electric lighting (fluorescent lamps, incandescent wire filaments, mercury vapor lamps, ultraviolet lamps), pesticides, pharmaceuticals (e.g., nasal sprays, cosmetics, contact lens products), household detergents and cleaners, laboratory chemicals, inks and paper coatings, lubrication oils, wiring devices and switches, and textiles. Though mercury use in many of the above items being produced now is restricted or banned, there are still some existing, older products in use [22].

Terrestrial plants are generally insensitive to the harmful effects of mercury compounds; however, mercury is known to affect photosynthesis and oxidative metabolism by interfering with electron transport in chloroplasts and mitochondria. Mercury also inhibits the activity of aquaporins and reduces plant water uptake [27].

Mercury and its compounds are cumulative toxins and in small quantities are hazardous to human health. The major effects of mercury poisoning manifest as neurological and renal disturbances as it can easily pass the blood-brain barrier and has effect on the brain [26].

PHYTOREMEDIATION TECHNOLOGY

Phytoremediation techniques have been briefly depicted in many literatures or articles. The generic term "phytoremediation" consists of the Greek prefix phyto (plant), attached to the Latin root remedium (to correct or remove an evil) [28, 29]. Some definitions on phytoremediation that have been described by several researchers are listed in Table 1.

Table 1: Definition of phytoremediation

No.	Researchers	Definition of phytoremediation
(1)	[30]	The use of plants to improve degraded environments
(2)	[31]	The use of plants, including trees and grasses, to remove, destroy or sequester hazardous contaminants from media such as air, water, and soil
(3)	[24]	The use of plants to remediate toxic chemicals found in contaminated soil, sludge, sediment, ground water, surface water, and wastewater
(4)	[32]	An emerging technology using specially selected and engineered metal accumulating plants for environmental cleanup
(5)	[33]	The use of vascular plants to remove pollutants from the environment or to render them harmless

(6)	[3]	The engineered use of green plant to remove, contain, or render harmless such environmental contaminants as heavy metals, trace elements, organic compounds, and radioactive compounds in soil or water. This definition includes all plant-influenced biological, chemical, and physical processes that aid in the uptake, sequestration, degradation, and metabolism of contaminants, either by plants or by the free-living organisms that constitute the plant rhizosphere
(7)	[29]	Phytoremediation is the name given to a set of technologies that use different plants as a containment, destruction, or an extraction technique. Phytoremediation is an emerging technology that uses various plants to degrade, extract, contain, or immobilize contaminants from soil and water
(8)	[34]	Phytoremediation in general implies the use of plants (in combination with their associated microorganisms) to remove, degrade, or stabilize contaminants

Generally, according to the above researchers, phytoremediation is defined as an emerging technology using selected plants to clean up the contaminated environment from hazardous contaminant to improve the environment quality. Figure 1 depicts the uptake mechanisms of both organics and inorganics contaminants through phytoremediation technology. For organics, it involves phytostabilization, rhizodegradation, rhizofiltration, phytodegradation, and phytovolatilization. These mechanisms related to organic contaminant property are not able to be absorbed into the plant tissue. For inorganics, mechanisms which can be involved are phytostabilization, rhizofiltration, phytoaccumulation and phytovolatilization.

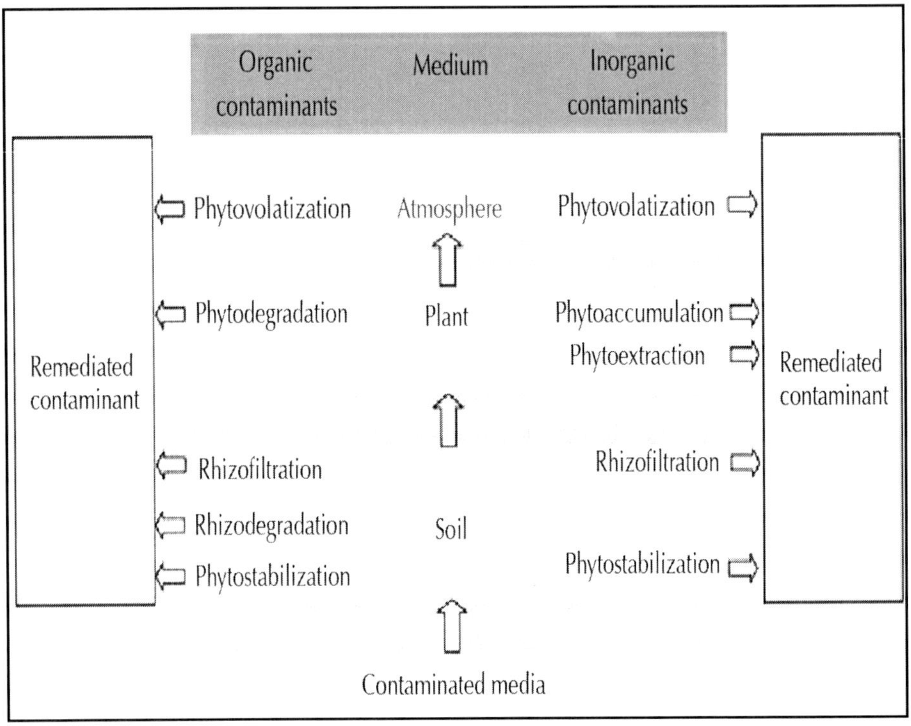

Figure 1: Uptake mechanisms on phytoremediation technology. Source: [35].

Based on Figure 1, some certain essential processes involved in phytoremediation technology [29, 31] are phytostabilization and phytoextraction for inorganic contaminants, and phytotransformation/phytodegradation, rhizofiltration, and rhizodegradation for organic contaminants.

The root plants exudates to stabilize, demobilize and bind the contaminants in the soil matrix, thereby reducing their bioavailability. These all are called as phytostabilization process. Certain plant species have used to immobilize contaminants in the soil and ground water through absorption and accumulation by roots, adsorption onto roots, or precipitation within the root zone. This process is for organics and metals contaminants in soils, sediments, and sludges medium [29, 31].

Specific plant species can absorb and hyperaccumulate metal contaminants and/or excess nutrients in harvestable root and shoot tissue, from the growth substrate through phytoextraction process. This is for metals, metalloids, radionuclides, nonmetals, and organics contaminants in soils, sediments, and sludges medium [29, 31].

Phytovolatilization process is the plants ability to absorb and subsequently volatilize the contaminant into the atmosphere. This process is for metal contaminants in groundwater, soils, sediments, and sludges medium. Since phytotransformation/phytodegradation process is the breakdown of contaminants taken up by plants through metabolic processes within the plant or the breakdown of contaminants externally to the plant through the effect of compounds produced by the plants. This process is for complex organic molecules that are degraded into simpler molecule contaminants in soils, sediments, sludges, and groundwater medium [29,31].

Plant roots take up metal contaminants and/or excess nutrients from growth substrates through rhizofiltration (=root) process, the adsorption, or, precipitation onto plant roots or absorption into the roots of contaminants that are in solution surrounding the root zone. This process is for metals, excess nutrients, and radionuclide contaminants in groundwater, surface water, and wastewater medium [29, 31].

The breakdown of contaminants in the soil through microbial activity that is enhanced by the presence of the root zone is called rhizodegradation. This process uses microorganisms to consume and digest organic substances for nutrition and energy. Natural substances released by the plant roots, sugars, alcohols, and acids, contain organic carbon that provides food for soil microorganisms and establish a dense root mass that takes up large quantities of water. This process is for organic substance contaminants in soil medium [29, 31].

MECHANISMS OF HEAVY METAL UPTAKE BY PLANT

Contaminant uptake by plants and its mechanisms have been being explored by several researchers. It could be used to optimize the factors to improve the performance of plant uptake. According to Sinha et al. [36], the plants act both as "accumulators" and "excluders". Accumulators survive despite concentrating contaminants in their aerial tissues. They biodegrade or biotransform the contaminants into inert forms in their tissues. The excluders restrict contaminant uptake into their biomass.

Plants have evolved highly specific and very efficient mechanisms to obtain essential micronutrients from the environment, even when present at low ppm levels. Plant roots, aided by plant-produced chelating agents and plant-induced pH changes and redox reactions, are able to solubilize and take up micronutrients from very low levels in the soil, even from nearly insoluble precipitates. Plants have also evolved highly specific mechanisms to translocate and store micronutrients. These same mechanisms are also involved in the uptake, translocation, and storage of toxic elements, whose chemical properties simulate those of essential elements. Thus, micronutrient uptake mechanisms are of great interest to phytoremediation [37].

The range of known transport mechanisms or specialized proteins embedded in the plant cell plasma membrane involved in ion uptake and translocation include (1) proton pumps ("-ATPases that consume energy and generate electrochemical gradients), (2) co- and antitransporters (proteins that use the electrochemical gradients generated by "-ATPases to drive the active uptake of ions), and (3) channels (proteins that facilitate the transport of ions into the cell). Each transport mechanism is likely to take up a range of ions. A basic problem is the interaction of ionic species during uptake of various heavy metal contaminants. After uptake by roots, translocation into shoots is desirable because the harvest of root biomass is generally not feasible. Little is known regarding the forms in which metal ions are transported from the roots to the shoots [37].

Plant uptake-translocation mechanisms are likely to be closely regulated. Plants generally do not accumulate trace elements beyond near-term metabolic needs. And these requirements are small ranging from 10 to 15 ppm of most trace elements suffice for most needs [37]. The exceptions are "hyperaccumulator" plants, which can take up toxic metal ions at levels in the thousands of ppm. Another issue is the form in which toxic metal ions are stored in plants, particularly in hyperaccumulating plants, and how these plants avoid metal toxicity. Multiple mechanisms are involved. Storage in the vacuole appears to be a major one [37].

Water, evaporating from plant leaves, serves as a pump to absorb nutrients and other soil substances into plant roots. This process, termed evapotranspiration, is responsible for moving contamination into the plant shoots as well. Since contamination is translocated from roots to the shoots, which are harvested, contamination is removed while leaving the original soil undisturbed. Some plants that are used in phytoextraction strategies are termed "hyperaccumulators." They are plants that achieve a shoot-to-root metal-concentration ratio greater than one. Nonaccumulating plants typically have a shoot-to-root ratio considerably less than one. Ideally, hyperaccumulators should thrive in toxic environments, require little maintenance and produce high biomass, although few plants perfectly fulfill these requirements [38].

Metal accumulating plant species can concentrate heavy metals like Cd, Zn, Co, Mn, Ni, and Pb up to 100 or 1000 times those taken up by nonaccumulator (excluder) plants. In most cases, microorganisms bacteria and fungi, living in the rhizosphere closely associated with plants, may contribute to mobilize metal ions, increasing the bioavailable fraction. Their role in eliminating organic contaminants is even more significant than that in case of inorganic compounds [39, 40].

Heavy metal uptake by plant through phytoremediation technologies is using these mechanisms of phytoextraction, phytostabilisation, rhizofiltration, and phytovolatilization as shown in Figure 2.

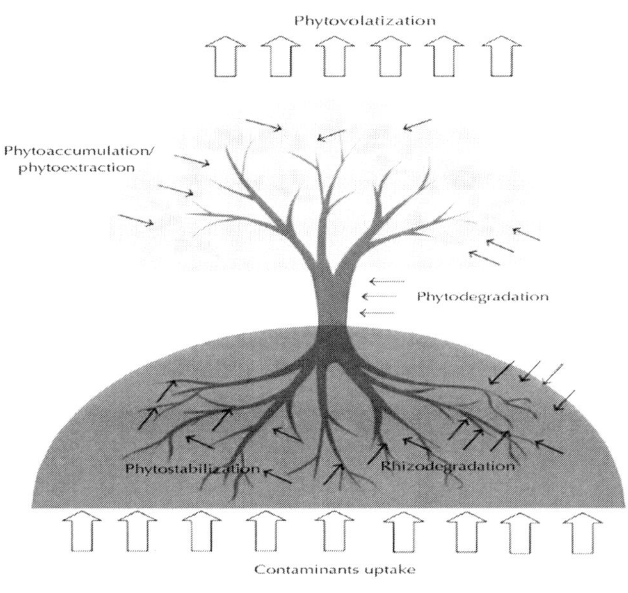

Figure 2: The mechanisms of heavy metals uptake by plant through phytoremediation technology.

Phytoextraction

Phytoextraction is the uptake/absorption and translocation of contaminants by plant roots into the above ground portions of the plants (shoots) that can be harvested and burned gaining energy and recycling the metal from the ash [28, 39–42].

Phytostabilisation

Phytostabilisation is the use of certain plant species to immobilize the contaminants in the soil and groundwater through absorption and accumulation in plant tissues, adsorption onto roots, or precipitation within the root zone preventing their migration in soil, as well as their movement by erosion and deflation [28,39–42].

Rhizofiltration

Rhizofiltration is the adsorption or precipitation onto plant roots or absorption into and sequesterization in the roots of contaminants that are in solution surrounding the root zone by constructed wetland for cleaning up communal wastewater [28, 39–42].

Phytovolatilization

Phytovolatilization is the uptake and transpiration of a contaminant by a plant, with release of the contaminant or a modified form of the contaminant to the atmosphere from the plant. Phytovolatilization occurs as growing trees and other plants take up water along with the contaminants. Some of these contaminants can pass through the plants to the leaves and volatilize into the atmosphere at comparatively low concentrations [28, 39–42].

Plants also perform an important secondary role in physically stabilizing the soil with their root system, preventing erosion, protecting the soil surface, and reducing the impact of rain. At the same time, plant roots release nutrients that sustain a rich microbial community in the rhizosphere. Bacterial community composition in the rhizosphere is affected by complex interactions between soil type, plant species, and root zone location. Microbial populations are generally higher in the rhizosphere than in the root-free soil. This is due to a symbiotic relationship between soil microorganisms and plants. This symbiotic relationship can enhance some bioremediation processes. Plant roots also may provide surfaces for sorption or precipitation of metal contaminants [27].

In phytoremediation, the root zone is of special interest. The contaminants can be absorbed by the root to be subsequently stored or metabolised by the plant. Degradation of contaminants in the soil by plant enzymes exuded from the roots is another phytoremediation mechanism [43].

For many contaminants, passive uptake via micropores in the root cell walls may be a major route into the root, where degradation can take place [3].

FACTORS AFFECTING THE UPTAKE MECHANISMS

There are several factors which can affect the uptake mechanism of heavy metals, as shown in Figure 3. By having knowledge about these factors, the uptake performance by plant can be greatly improved.

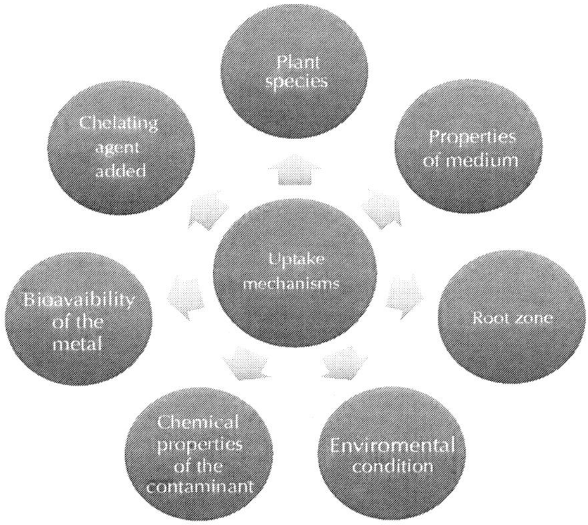

Figure 3: Factors which are affecting the uptake mechanisms of heavy metals.

The Plant Species

Plants species or varieties are screened, and those with superior remediation properties are selected [31]. The uptake of a compound is affected by plant species characteristic [44]. The success of the phytoextraction technique depends upon the identification of suitable plant species that hyperaccumulate heavy metals and produce large amounts of biomass using established crop production and management practices [24].

Properties of Medium

Agronomical practices are developed to enhance remediation (pH adjustment, addition of chelators, fertilizers) [31]. For example, the amount of lead absorbed by plants is affected by the pH, organic matter, and the phosphorus content of the soil. To reduce lead uptake by plants, the pH of the soil is adjusted with lime to a level of 6.5 to 7.0 [20].

The Root Zone

The Root Zone is of special interest in phytoremediation. It can absorb contaminants and store or metabolize it inside the plant tissue. Degradation of contaminants in the soil by plant enzymes exuded from the roots is another phytoremediation mechanism. A morphological adaptation to drought stress is an increase in root diameter and reduced root elongation as a response to less permeability of the dried soil [43].

Vegetative Uptake

Vegetative Uptake is affected by the environmental conditions [44]. The temperature affects growth substances and consequently root length. Root structure under field conditions differs from that under greenhouse condition [43]. The success of phytoremediation, more specifically phytoextraction, depends on a contaminant-specific hyperaccumulator [45]. Understanding mass balance analyses and the metabolic fate of pollutants in plants are the keys to proving the applicability of phytoremediation [46].

Metal uptake by plants depends on the bioavailability of the metal in the water phase, which in turn depends on the retention time of the metal, as well as the interaction with other elements and substances in the water. Furthermore, when metals have been bound to the soil, the pH, redox potential, and organic matter content will all affect the tendency of the metal to exist in ionic and plant-available form. Plants will affect the soil through their ability to lower the pH and oxygenate the sediment, which affects the availability of the metals [47], increasing the bioavailability of heavy metals by the addition of biodegradable physicochemical factors, such as chelating agents and micronutrients [34].

Addition of Chelating Agent

The increase of the uptake of heavy metals by the energy crops can be influenced by increasing the bioavailability of heavy metals through addition of biodegradable physicochemical factors such as chelating agents, and micronutrients, and also by stimulating the heavy-metal-uptake capacity of the microbial community in and around the plant. This faster uptake of heavy metals will result in shorter and, therefore, less expensive remediation periods. However, with the use of synthetic chelating agents, the risk of increased leaching must be taken into account [34]. The use of chelating agents in heavy-metal-contaminated soils could promote leaching of the contaminants into the soil. Since the bioavailability of heavy metals in soils decreases above pH 5.5–6, the use of a chelating agent is warranted, and may be required, in alkaline soils. It was found that exposing plants to EDTA for a longer period (2 weeks) could improve metal translocation in plant tissue as well as the overall phytoextraction performance. The application of a synthetic chelating agent (EDTA) at 5 mmol/kg yielded positive results [8]. Plant roots exude organic acids such as citrate and oxalate, which affect the bioavailability of metals. In chelate-assisted phytoremediation, synthetic chelating agents such as NTA and EDTA are added to enhance the phytoextraction of soil-polluting heavy metals. The presence of a ligand affects the biouptake of heavy metals through the formation of metal-ligand complexes and changes the potential to leach metals below the root zone [48].

EFFECTIVENESS OF HEAVY METALS UPTAKE BY PLANTS

Several studies have described the performance of heavy metals uptake by plants. It is reported that phytoremediation technology is an alternative to treat heavy-metal-contaminated side which will be more admitted in order to remediate the environment. Table 2 lists some research done to remediate heavy metals from contaminated soil, while Table 3 lists some research conducted to remediate them from contaminated water and wastewater.

Table 2: Phytoremediation study on soil medium

No.	Researcher	Research scale and duration	Uptake mechanisms and media (substrate)	Contaminant or parameter and concentration	Plants name and type	Result
(1)	[16]	Greenhouse pot experiment (6, 10, and 16 days)	Phytoextraction (soil was added to aqueous solution and was dried overnight in an oven at 120°C, cooled, and transferred to the pot)	Aqueous solution containing 0.1041 g of sodium arsenate heptahydrate (Na2HAsO4·7H2O), the mixture which contained 50 mg/kg of As (wet weight)	Leersia oryzoides(rice-cut grass)—terrestrial plant	The increase in plant size is matched by a decrease in shoot arsenic concentration. The data show that 12, 13, and 13 mg/m2 of arsenic were absorbed by the shoots at 6, 10, and 16 weeks, respectively. Since the SRQ and PECs all exhibit the same downward trend after 6 wk, it is suggested that periodic mowing of Leersia oryzoidesgrown for phytoextraction purposes on contaminated land could maintain the high arsenic uptake at 6 week.
(2)	[33]	Laboratory (pot experiment) (90 days)	Fly ash and soil mixtures	Pb as lead nitrate, Zn as zinc sulfate, Ni as nickel sulfate, Mn as manganese chloride, and Cu as copper sulfate (1000 ppm concentration each (Spiked))	Scirpus littoralis—semiaquatic	The metal content ratios BO/soil (B/S) were higher than shoot/soil ratios (T/S) for all the metals, the highest being for Ni. Metal ratios BO/water (B/W) were also higher than shoot/water (T/W) ratios, but the B/W ratio was maximum for Zn. All the metals except Ni showed negative correlation with nitrogen but they were all nonsignificant. However, Puptake showed positive correlations with all the metals, and all were significant at 1% confidence limit.

(3)	[49]	Field study (90 days)	Soil (agricultural land area)	(Cu, Cd, Cr, Zn, Fe, Ni, Mn, and Pb)	Wheat (Triticum aestivum L.)—terrestrial Indian mustard (Brassica campestris L.)—terrestrial	Analyses of effluents and soil samples have shown high metal content than the permissible limit except Pb. Analyses of plant samples have indicated the maximum accumulation of Fe followed by Mn and Zn in root > shoot > leaves > seeds. Maximum increase in photosynthetic pigment was observed between 30 and 60 days while protein content was found maximum between 60 and 90 days of growth period in both plants.
(4)	[5]	Laboratory (65 days)	Phytoextraction (soil)	Pb by using standard Pb solutions (75 mg Pb/1 kg soil)	Creeping zinnia (Alternanthera phyloxeroides)—aquatic Moss rose (Sanvitalia procumbens)—terrestrial Alligator weed (Portulaca grandiflora)—aquatic	Alternanthera phyloxeroides shows the highest lead content in its tissues. This might be caused by it forming long stolons, a massive fibrous root system, and large surface area which benefits the accumulation of lead. Efficiency process 30–80%.
(5)	[34]	Literature review	Soil	Cd, Cr, Cu, Ni, Pb, and Zn	Brassica juncea(Indian mustard),Brassica rapa (field mustard), andBrassica napus(rape)—terrestrial	Brassica rapa exhibited the highest affinity for accumulating Cd and Pb from the soil, either with/without additional use of mobilizing soil amendments. Two Brassica species (Brassica napus and Raphanus sativus) were moderately tolerant when grown on a multi-metalcontaminated soil. The distribution of heavy metals in the organs of crops decreased in the following order: leaves > stems > roots > fruit shell > seeds.

(6)	[50]	Laboratory—pot experiment (12 days)	Agropeat and half strength Hoagland solution	Arsenic (As) as of sodium (meta-) arsenite (50 uM, 150 uM and 300 uM)	Brassica juncea var. Varuna and Pusa Bold—terrestrial	Increase/decrease of antioxidant enzymes activities showed not much changes at the given concentrations. The data presented indicates the differential responses in both the varieties and also that the increased tolerance in P. Bold may be due to the defensive role of antioxidant enzymes, induction of MAPK, and upregulation of PCS transcript which is responsible for the production of metal-binding peptides.
(7)	[51]	Field study (trials to extract heavy metals from two contaminated soils, one calcareous (5 years) and one acidic (2 years))	Phytoextraction (soil)	Cd and Zn	Willow (Salix viminalis)—terrestrial	Salix had performed better on the acidic soil because of larger biomass production and higher metal concentrations in shoots. Addition of elemental sulphur to the soil did not yield any additional benefit in the long term, but application of an Fe chelate improved the biomass production. Cd and Zn concentrations were significantly higher in leaves than stems. On both soils, concentration in shoots decreased with time.
(8)	[52]	Laboratory (26 days)	Sludge-amended soils	Cd and Zn	Raphanus sativus L.	This study has shown that clear evidence of as ludge-driven plateau response in metal uptake by plants will only be obtained when studies have found a good hyperbolic relationship between soil solution metal concentration with increasing sludge application rate and can link this to a plateau response in plant uptake of metals.

		Laboratory—lysimeter pot (March 1995–September 1995)	Soil	Zn as ZnSO4(50, 1,500, 2,000 μg/g (ppm) Zn. and 2,000 μg/g (ppm), and 0 μg/g (ppm) (control) received nutrient only)	Hybrid poplar (Populus sp.)—terrestrial	At levels of zinc above 1,000 μg/g (ppm) in nutrient added, leachate levels were always below 100 μg/g (ppm) in samples as the zinc addition; these levels increased the following day and then decreased sharply the second day after the zinc addition, to concentrations less than 100 μg/g (ppm). The zinc concentration steadily decreased as the plants apparently reabsorbed the zinc as the nutrient was cycled through the pots on subsequent days. The root tissues showed much higher concentrations of accumulated and sequestered metal than did the above ground parts.
(9)	[3]	Laboratory (April 1996, 2 months)	Soil	Zn (160 μg/g Zn, 600 μg/g Zn, and 0 μg/g Zn (control))	Eastern gamagrass (Tripsacum dactyloides)—terrestrial	Leachate analyses for zinc indicate that initially plants subjected to both levels of zinc were removing up to 70% of the zinc from the leachate. The plants receiving 160 μg/g Zn had grown considerably and were almost the same size as the controls (no zinc), but some of the mature leaf blades were rolled; the mean zinc removal rate for these plants was 50% of the zinc in the leachate. The plants receiving 600 μg/g Zn were smaller than the controls, their color was a darker green, most of the mature leaf blades were rolled, and the mean zinc removal rate was about 30% of the zinc in the leachate.

			Soil	Pb and As (up to 1000 μg/g Pb and up to 200 μg/g As)	Hybrid willow (Salix sp.) and hybrid poplar (Populus sp.)—terrestrial	The willows were able to remove approximately 9.5% of the available lead and about 1% of the total arsenic from the contaminated soil. The less mature poplars removed about 1% of the available lead and 0.1% of the total arsenic from the same soil. In the sand experiment, the willows took up about 40% of the administered lead and 30 to 40% of the administered arsenic.
(10)	[53]	Field (1976–2001)	(Soil)	Nonessential (Cd, Ni, Pb) and essential heavy metals (Cu, Fe, Mn, Zn). The tetrasodium salt of EDTA was applied at rates of 0, 0.5, 1, 2 g EDTA salt/kg surface (25 cm depth) soil	Sunflower (Helianthus annuusL.) and Hybrid poplar (Populus deltoides Marsh. x P. nigra L.)—terrestrial	For sunflower, the 1.0 g/kg rate of chelate addition resulted in maximal removal of the three nonessential heavy metals (Cd, Ni, Pb). Uptake of the essential heavy metals by sunflower was little affected by the EDTA. The leaves of sunflower grown with 1.0 g EDTA $Na4\cdot 2H2O$/kg soil accumulated more Cd, Ni, and Pb than leaves of sunflower grown without the EDTA salt. Removal of the non-essential heavy metals by sunflower was greater at the higher plant density compared to the lower one.

(11)	[54]	Laboratory	18 different phytoremedation treatments. I. parcel: mine waste without fly ash. Control and untreated plot. 3 test plants. II. mine waste + fly ash without liming. Control and untreated plot. 3 test plants. III. mine waste + fly ash + liming. Control and untreated plot. 3 test plants.	As, Cd, Mo, Pb, Zn (soil) and As, Cd, Pb, Ni, Zn (water)	Grasses (mixture of selected species), sorghum (Sorghum bicolor L.) and Sudan grass (Sorghum sudanense)—terrestrial	The chemical risks of the Gyöngyösoroszi spoils were assessed. The major contaminants of the waste mine were identified: Pb, Zn, Cd, As. The concept of the integrated phytoremediation was successfully applied to vegetate Gyöngyösoroszi spoil. The biomass production was different, depending on the technology variant. The highest biomass production was achieved, when multilevel revitalization was also applied. The integrated phytoremediation treatments not only produced high biomass, but also decreased the heavy metal content in the plants.
(12)	[55]	Field (1995–1997)	Soil	Ni, Cu, Cd, Zn	Willow (Salix spp.)—terrestrial	One group of willow had relatively low Ni and Cu in the bark and high Cd and Zn in the wood, with a good survival rate and biomass production. The second group of willow had relatively high Ni and Cu in the bark and low Cd and Zn in the wood and performed poorly in terms of survival and biomass production.

(13)	[8]	Laboratory (15 May and 25 September 2002)	Phytoextraction (soil)	Cu, Pb, Zn	Fescue (Festuca arundinaceaSchreb.), Indian mustard (Brassica juncea (L.) Czern.), and willow (Salix viminalis L.)—terrestrial	The use of the freeacid form of EDTA and exposure time of one to two weeks before harvesting increased the concentration of metals translocated to plant tissues. It is found no significant difference in heavy metal concentrations in higher and lower soil horizons between EDTA treated and untreated soils. Exposing plants to EDTA for a longer period (2 weeks) could improve metal translocation in plant tissue as well as the overall phytoextraction performance.
(14)	[24]	Field experiment (3 years)	Phytoextraction (soil content with Hg)	Hg (mean Hg content of the soil was 29.17 μg/g for the 0–10 cm horizon and 20.32 μg/g for 10–40 cm horizon with less than 2% of the total Hg being bioavailable)	Three agriculture crop plants:Triticum aestivum(wheat)—terrestrial Hordeum vulgare(barley)—terrestrial Lupinus luteus(yellow lupin)—terrestrial	The decrease of mean Hg concentration from 29.17 μg g–1 at 0–10 cm horizon to 20.32 μg g–1 at 10–40 cm horizon demonstrated the anthropogenic origin of the mercury in the soil. Preliminary results show that all crops extracted mercury, with Hg plant concentration reaching up to 0.479 μg g–1 in wheat. The mercury concentration in the plants accounted for less than 3% of mercury concentration in the soil. The Hg concentrations in the plants were similar or even higher than that of the bioavailable Hg in the soils. Mercury extraction yields reached up to 719 mg/ha for barley.

(15)	[56]	Pot experiment (20 weeks)	Soil from waste deposits of the lead smelter	Pb	Agrostis capillaris—terrestrial	Inoculation with indigenous or nonindigenous AMF in this experiment did not decrease Pb uptake by the host in comparison with nonmycorrhizal plants grown in contaminated soil. It can be concluded that 13 months of subculturing in an inert substrate did not affect development of G. intraradicesPH5 isolated from the waste deposits of a Pb smelter in contaminated soil of its origin. The interaction of the fungus with the host plant was changed: the ability of the lineage cultured without HM to support plant growth in Pb-contaminated soil was decreased, while translocation of Pb from plant roots to shoots increased.
(16)	[38]	Field and greenhouse experiments	Phytoextraction (As- and Pb-contaminated soil)	Arsenic (As) and lead (Pb)	Chinese Brake Ferns (Pteris vittata)—terrestrial Indian Mustard (Brassica juncea)—terrestrial	It appears that EDTA is necessary for Pb extraction due to the low soil Pb bioavailability. Soil amendments like EDTA are necessary because they mobilize soil Pb, making it available to plant roots. It may not be advisable to apply EDTA in the environment, because EDTA mobilizes metals, which may leach into surrounding property of groundwater. The presence of other metals that compete for EDTA may increase the amount of EDTA required for Pb remediation.

(17)	[27]	Laboratory experiment, using chamber (6 weeks)	Phytostabilization (mercury-contaminated soil used in this experiments was obtained from a chemical factory located in the southeast part of Poland, which has been in operation for over 50 years)	Hg	Species Festuca rubra (red fescue)—terrestrial Poa pratensis (meadow grass)—terrestrialArmoracia lapathifolia(horseradish)—terrestrial Helianthus tuberosus(Jerusalem sunflower)—terrestrial S. viminalis(willow)—terrestrial	The highest concentrations of mercury were found at the roots, but translocation to the aerial part also occurred. Most of the plant species tested displayed good growth on mercury contaminated soil and sustained a rich microbial population in the rhizosphere. An inverse correlation between the number of sulfur amino acid decomposing bacteria, and root mercury content was observed. These results indicate the potential for using some species of plants to treat mercury-contaminated soil through stabilization rather than extraction.
(18)	[57]	Field (July and October)	Phytoextraction and phytostabilisation (soil)	Zn, Cu, Cr and Cd	Two poplar clones (Populus deltoides x maximowiczii-clone Eridano andP. x euramericana-clone I-214)—terrestrial	Leaf, stem, root and woody cutting biomasses of treated plants were significantly greater than those in the controls in both clones, except for stem biomass at the beginning of October. Among the four heavy metals (Zn, Cu, Cr, and Cd), only Zn, Cu, and Cr concentrations in plants differed consistently between clones or soil treatments, while Cd levels were always below the detection limits.

(19)	[58]	Field study and laboratory experiment (2002-2003 (field study), 3 months for laboratory experiment)	Soil	Fe, Zn, Pb, Cu, Ni, Cr, Mn	Brachythecium populeum	The results obtained from this study onB. populeum lead to the inference that physiological/bio-chemical analysis of epiphytic bryophytes can serve as cost-effective indicators/ monitors for the environmental quality of any area, and on the basis of this information appropriate steps can be taken to improve the air quality of an area.
(20)	[59]	Pot experiment and field trial (2004-2005 for pot experiment, and 2005 field trial)	Phytoextraction and phytostabilization (soil)	As, Co, Cu, Pb, and Zn	Three poplar species (Populus alba, Populus nigra, Populus tremula) and Salix alba—terrestrial	Trace element concentrations were much higher in roots than in above-ground tissues, with particularly high concentrations in fine roots. The highest accumulations were measured in P. nigraand S. alba. In wood, the highest concentrations of Cu and Zn were in S. alba. Salix alba foliage contained highest concentrations of As, Cu, Pb, and Zn; leaf Zn concentration exceeded those of wood by almost 6 times. The overall removal of trace elements was only significantly higher in P. alba than in S. alba; P. alba.

(21)	[60]	Pot experiment and field trial (2 years (2004-2005) for pot experiment and field trial on May–September 2005)	Phytoextraction and phytostabilization (soil (Pyrite ore contains mainly pyrite (FeS_2), lesser amounts of chalcopyrite ($CuFeS_2$), sphalerite (ZnS), magnetite (Fe_3O_4), and various trace elements))	As, Co, Cu, Pb and Zn	P. alba L. (white poplar)—terrestrial P. nigra L. (black poplar)—terrestrial P. tremula L. (European aspen)—terrestrial Salix alba L. (white willow)—terrestrial	The result shown that establishment ofPopulus and Salix species at the site is achievable through ripping of the surface, minimal tillage, some mixing of the wastes with imported soil, irrigation and fertilisers. Potentially, the elevated concentrations of Pb, As and other elements could be leached from the remediated wastes towards groundwater or other receptors, and these fluxes could also be influenced by soil amendments, changes in the rhizosphere or both. Immobilisation of trace elements in both coarse and fine roots may reduce leaching, particularly of Cu and Zn but also As and Pb.
(22)	[61]	Greenhouse	Phytoextraction and phytostabilization (soil)	Six sediment-derived soils with increasing field Cd levels (0.9–41.4 mg/kg)	Two willow clones (Salix fragilis "Belgisch Rood" andSalix viminalis "Aage")—terrestrial	No growth inhibition was observed for both clones for any of the treatments. Dry weight root biomass and total shoot length were significantly lower for S. viminalis compared to S. fragilis for all treatments. Willow foliar Cd concentrations were strongly correlated with soil and soil water Cd concentrations. Both clones exhibited high accumulation levels of Cd and Zn in above ground plant parts. Cu, Cr, Pb, Fe, Mn, and Ni were found mainly in the roots. Bioconcentration factors of Cd and Zn in the leaves were the highest for the treatments with the lowest soil Cd and Zn concentration.

(23)	[62]	Laboratory and field	Rhizobox experiment was used to investigate the short-term effect of willow roots on metal availability in oxic and anoxic sediment. Longer-term effects were assessed in a field trial (soil)	Cd, Zn, Cu, and Pb	Willow (Salix spp.)—terrestrial	The rhizobox trial showed that Cd, Zn, and Cu extractability in the rhizosphere increased while the opposite was observed for Pb. The field trial showed that Cu and Pb, but not Cd, were more available in the root zone after water and ammonium acetate (pH 7) extraction compared with the bulk sediment. Sediment in the root zone was better structured and aggregated and thus more permeable for downward water flows, causing leaching of a fraction of the metals and significantly lower total contents of Cd, Cu, and Pb.
(24)	[63]	Pot experiment	Phytoextraction (soil)	As (as $Na2HAsO4$), Cd (as $CdCl2$), Pb (as $Pb(CH3COO)2$), and Zn (as $Zn(CH3COO)2$) (100 mg As/kg, 40 mg Cd/kg, 2000 mg Pb/kg, and 2000 mg Zn/kg)	Salix spp.—terrestrial	Although As and Cd uptake slightly increased in Suchdol-Zn soil compared to Suchdol-Pb soil, the element removal from soil was significantly higher in Suchdol-Pb soil due to a significant reduction of aboveground biomass yield in Suchdol-Zn soil. The yield reduction decreased the uptake of plant-available elements by biomass; thus higher plant-available portions of As and Cd were found in Suchdol-Zn soil.

(25)	[64]	Field survey: from 12 As-contaminated sites (September to November 2003)	Field study: contaminated soil	As	Samples of 24 fern species belonging to 16 genera and 11 families as well as their associated soils were collected—terrestrial	Pteris multifida and P. oshimensis can (hyper-) accumulate As in their fronds with high concentrations. Total As concentrations in soils associated with P. multifida and P. oshimensis varied from 1262 to 47,235 mg/kg, but the DTPA-extractable As concentrations were relatively low, with a maximum of 65 mg/kg. Although As concentrations in the fronds of P. oshimensis were comparatively lower than those of P. multifida, its high above ground biomass makes it more suitable for phytoremediating As-contaminated soils.

(26)	[65]	Field survey (contaminated site since 1976; the sample was taken in 2006)	Soil	Cu, Pb, Cd, and Zn	Paulowni fortunei(seem) Hems	In the rhizosphere and bulk soils of P. fortunei, all physico-chemical properties increased with the revegetation time. The total contents of Cu, Pb, Cd, and Zn also consistently increased with the re-vegetation time; moreover, rhizosphere soils accumulated more heavy metals than bulk soils with the revegetation time. In the rhizosphere soils of P. fortunel, the immobility and bioavailability of heavy metals were enhanced. In the rhizosphere microenvironment, pH, OM, and EC were important factors affecting the distribution of heavy metal fractions. Among different heavy metal fractions, the exchangeable and organically bound fractions were easily available for P. fortunei, but carbonate, Fe–Mn oxide, and residual fractions were not easily available for P. fortunei.

(27)	[4]	Greenhouse pot experiment (August-September 2002)	Soil was sampled in two sites: contaminated soil was taken near road with heavy traffic and clean soil was taken from park protected from the road by buildings	Ag, As, B, Ba, Be, Bi, Ca, Cd, Co, Cr, Cu, Fe, K, Li, Mg, Mn, Mo, Na, Ni, P, Pb, Rb, S, Sb, Se, Sr, Th, Ti, Tl, U, V and Zn	Wheat Triticum vulgare, sort Umanka—terrestrial	Concentrations of Ag, Cd, Cu, Pb, Sb, and Zn in the initial contaminated soil were 3–6 times higher than those in the initial clean soil. In particular, contents of Cu, Mo, Ni, Pb, Sb and Zn in roots of the wheat grown in the contaminated soil were higher than those in the roots of the plants grown in the clean soil. Moreover, all the elements except Pb transferred more easily from roots to leaves.
(28)	[66]	Field experiment (155 days (May–November))	Soil (agricultural soil)	Cd, Cr, Pb, As, and Hg	Rice (Oryza sativaL.)—terrestrial	The results showed the rice grain contained significantly lower amounts of five metals than straw and root in all sampling sites. Rice root accumulated Cd, As, and Hg from the paddy soil. The rice plant transported As very weakly, whereas Hg was transported most easily into the straw and grain among studied heavy metals.

No. (7), (10), (12), (18), (22), (23) adapted from no. (20). Phytoremediation Bibliography, Annotated Bibliography on Phytoremediation prepared by Mark Coleman, Biological Scientist, USDA Forest Service Southern Research Station and Ronald S. Zalesny Jr., Research Plant Geneticist, USDA Forest Service North Central Research Station May 1, 2006.

Table 3: Phytoremediation study on water medium (hydroponic)

No.	Researcher	Research scale and duration	Uptake mechanisms and media (substrate)	Contaminant or parameter and concentration	Plants	Result
(1)	[67]	Field study (October–July 2005)	Water of Tasik Chini	Cd, Cu, and Pb	Five aquatic plant species, Lepironia articulata, Pandanus helicopus, Scirpus grossus, Cabomba furcata, andNelumbo nucifera—aquatic	The highest concentration of heavy metals among the aquatic plants and plant parts was found in the roots of S. grossus. The concentrations of Cd in the leaves and stems of submerged aquatic plant, C. furcata, were higher than that in the leaves and stems of emergent aquatic plant and floating leaf plant. The concentration of Cu in the stem of C. furcata was greater than that in the leaf, while the concentration of Cd was more in the leaf than that in the stem. The heavy metal contents of the aquatic plants were in descending order of Pb > Cu > Cd. The highest internal translocation was found in P. helicopus, while the lowest internal translocation was found in S. grossus.

(2)	[68]	Laboratory (pot experiment)/14 days	Hydroponic	As and Se as $Na_2HAsO_4 \cdot 7H_2O$ and Na_2SeO_3/0, 0.73, 2.5, 4.27, 5.00 mg/L	Chinese brake fern (Pteris vittata L.)—terrestrial	At low levels of Se, As enhanced both Se uptake and the translocation of Se from roots to fronds. At higher levels of Se, As suppressed the uptake of Se. These results suggest that As serves to both stimulate and suppress Se uptake. The result is also in agreement with the well-known fact that Se is an element with both beneficial and toxic properties. The effect can change from beneficial to toxic based on the concentration of Se in plants.

(3)	[47]	Laboratory and Field study: wetland-pond system (Laboratory scale: 3 days cultivative and 84 hours exposure)	Field study: contaminated soil. Medium of laboratory scale experiment: L 0.1% Hoagland solution	Field study: Zn, Cu, Cd, and Pb. Laboratory: ZnCl2, CuCl2, CdCl2, and Pb(NO3)2 (mixture of 20 μmol Zn, 0.5 μmol Cu, 1.5 μmol Cd, and 1.5 μmol Pb/L)	Potamogeton natans L.-aquatic Lemna gibba L.-aquatic Alisma plantago-aquatica L.-aquatic Sagittaria sagittifolia L.-aquatic Juncus effusus L-aquaticLemna minor L.-aquatic Elodea canadensis Michx.-aquatic Lythrum salicariaL.-aquaticPhalaris arundinacea L.-aquatic Impatiens parviflora DC.-terrestrial Urtica dioica L.—terrestrial Filipendula ulmaria L.-aquatic P. natans-aquatic A. plantago-aquatica-aquatic F. ulmarina-aquatic	The aquatic plants seem to have a higher metal accumulation capacity in shoots than terrestrial plants. This may be due to the capacity of aquatic plants to take up by shoot directly from the water. When submersed and free-floating plants are actively growing and accumulating metals directly from the water, they will function as an effective filter in stormwater treatment. Emergent plants in general mediate the binding of these metals in the sediment. Also, the terrestrial plants have the capacity to bind Cd and Zn to their roots, and; therefore, they can mediate a good stabilization of these metals in soil.
(4)	[32]	Laboratory (15 days)	Hydroponic	Pb as (Pb(NO3)2)	Indian mustard (Brassica juncea var. megarrhiza)—terrestrial	Brassica juncea is one plant which accumulates high levels of Pb and other heavy metals. The results indicate that lead nitrate obviously inhibits the root, hypocotyls, and shoot growth ofBrassica juncea at the concentration of 10–3 M Pb2+. Brassica juncea has the ability to accumulate Pb primarily in its roots, transport, and concentrate it in its hypocotyls and shoots in much lesser concentrations.

(5)	[30]	Laboratory (5 days)	Phytofiltration (water)	Mercury as HgCl2 (0, 0.05, 0.5, 1, 2.5, 5, 10 mg/L)	Indian mustard (Brassica juncea)—terrestrial	Roots-concentrated Hg 100–270 times (on a dry weight basis) above initial solution concentrations. Mercury was more toxic to plants at 5 and 10 mg/L. The plants translocated little Hg to the shoots, which accounted for just 0.7–2% of the total Hg in the plants. Most Hg volatilisation occurred from the roots. Volatilised Hg was predominantly in the Hg(0) vapour form. Volatilisation was dependant on root uptake and absorption of Hg from the ambient solution. Efficiency process >95%.
(6)	[69]	Laboratory	Hydroponic	Arsenate (As(V)) and dimethylarsinic acid (DMAA)	Duckweed (Spirodela polyrhiza L.)—aquatic	The results show that not only internalized, but also surface-adsorbed arsenic (mostly arsenate) contributes significantly to the total amount of arsenic uptake in aquatic macrophyte S. polyrhiza L. The arsenic uptake in S. polyrhiza L. occurred through the phosphate uptake pathway as well as by physicochemical adsorption on Fe plaques of plant's surfaces. The arsenate uptake in the plant is related to the Fe ion and phosphate concentrations in culture medium while DMAA was not.

(7)	[2]	Laboratory (the contact times of 25–200 min were selected for the metal solutions (Co = 1.00 mM) with 2.0 g biomass/l at the obtained optimal pHs for each metal ion from the previous study)	Adsorption (water)	The Hg2+, Cr3+, Cr6+and Cu2+ stock solutions were prepared by dissolving their corresponding salts, viz. HgCl2, CrCl3·3H2O, K2Cr2O7, CuCl2(analytical grade from Merck) in distilled water (pH values were almost 7.0, 5.0, 3.0 and 6.0 for Hg2+, Cr3+, Cr6+ and Cu2+, respectively)	Lemna minor—aquatic	The potentiometric titration can be useful to study the pretreatment process of biomass (L. minor) using the acidic and alkali agents, the Qmax and KL values to remove Hg(II), Cr(III), Cr(VI), and Cu(II) from the aqueous solution by the activated L. minor at the alkali solution and by CaCl2/MgCl2/NaCl with 1 : 1 : 1 molar ratio were higher than those for the reference one at the same conditions, the removal percents of metal ions by no. ACS L. minor was higher than ACS one at the pre-treatment pHs before 7.0, but it was higher by ACS biomass than no. ACS one at the pre-treatment pHs after 7.0.
(8)	[70]	Laboratory (seedling 2 weeks and treatment 2 weeks)	Hydroponic	Hg and Au (0, 50, 100, and 200 uMHg (as Hg (CH3COO)2) and 0 and 50 uMAu (as KAuCl4) in hydroponics)	Chilopsis linearis(Cav.) sweet—terrestrial	The data showed that Au equimolar to Hg reduced the Hg toxicity. The concentration of Au and Hg in shoots indicated that C. linearis absorbed and translocated both Au and Hg at higher concentrations, compared to reported data. The data showed that the treatments produced structural alterations in both the vascular cylinder and the cortex. At the highest concentration, Hg produced a breakdown of the spongy parenchyma.

(9)	[71]	Laboratory (30 days)	Phytoextraction (water)	Mercury as $HgSO_4$ (0, 0.5 and 2 mg/L)	Water hyacinth (Eichornia crassipes)—aquatic Water lettuce (Pistia stratiotes)—aquatic Zebra rush (Scirpus tabernaemontani)—semi aquatic Taro (Colocasia esculenta)—aquatic	The higher the mercury concentration, the greater the amount of mercury removed by the plants. The largest uptake and accumulation capability is for water lettuce, followed by water hyacinth, taro and rush, respectively.
(10)	[72]	Laboratory—(pot experiment (10 days))	Hydroponic	As and Se (0, 150, or 300 uM of arsenat ($Na_2HAsO_4 \cdot 7H_2O$) in the presence of 0, 5 or 10 uM of selenat (Na_2SeO_4))	Pteris vitatta L.—terrestrial	Application of 5 uM Se enhanced As concentration by P. vittata fronds by 7–45%. At 5 uM, Se acted as an antioxidant, inhibiting lipid peroxidation (reduced by 26–42% in the fronds) via increased levels of thiols and glutathione (increased by 24% in the fronds). The results suggest that Se is either an antioxidant, or it activates plant protective mechanisms, thereby alleviating oxidative stress and improving arsenic uptake in P. vittata.

(11)	[45]	Laboratory (72 hours (for kinetics of Arsenic uptake), 3 days (effects of plant density, Plant re-use, and plant age), 10 days (groundwater remediation))	The groundwater was collected from a location which may have been contaminated from application of arsenical herbicides in the past.	As (pH 7.0, total As of 46 μg/L, As^{3+} of 1.6 μg/L, and total P of 20 μg/L)	Chinese Brake fern (Pteris vittata L.) plants—terrestrial	Chinese brake fern was efficient in taking up arsenic from a contaminated groundwater and was capable of reducing arsenic concentrations in the groundwater. One plant was sufficient to reduce arsenic in 600 mL groundwater to below 10 μg/L in 3 days. Young fern plants were more effective in arsenic removal than old fern plants of similar size. Ferns can be reused to remove arsenic from groundwater, but at a slower rate given the interval between exposures and nutritional status.
(12)	[73]	Laboratory	Hydroponic	Cu and Ni	Salix viminalisclones and the basket willow Black Maul (S. triandra). S. burjatica "Germany", S.x dasyclados, S. candida and S. spaethii—terrestrial	The more resistant clones produced more biomass in the glasshouse and field and had higher metal concentrations in the wood. The less resistant clones had greater concentrations of Cu and Ni in the bark and produced less biomass in the glasshouse and field. Significant relationships were found between the response of the same clones grown in the short-term glasshouse hydroponics system and in the field.

(13)	[74]	Laboratory (10 days cultivate and 7 days exposure)	Nutrient solution	As (0, 5, 10, 20, 40 and 80 uM)	Azolla: A. caroliniana andA. filiculoides—aquatic	The efflux of arsenate was much higher (by about 9-fold) than that of arsenite. This may be because most of arsenite inside the cells was complexed with thiol compounds. The high As-accumulating Azolla (A.caroliniana) released approximately two times more As than the low-As accumulating Azolla (A. filiculoides). It appears that the amount of As efflux was proportional to the amount of As accumulation in the two strains of Azolla.

No. (12) adapted from no. (20). Phytoremediation Bibliography, Annotated Bibliography on Phytoremediation prepared by Mark Coleman, Biological Scientist, USDA Forest Service Southern Research Station and Ronald S. Zalesny Jr., Research Plant Geneticist, USDA Forest Service North Central Research Station, May 1, 2006.

Based on the collected data from the phytoremediation research listed in Tables 2 and 3, the accumulation of heavy metals As, Pb, and Hg in plant tissue is summarized in respective, Figures 4, 5, and 6.

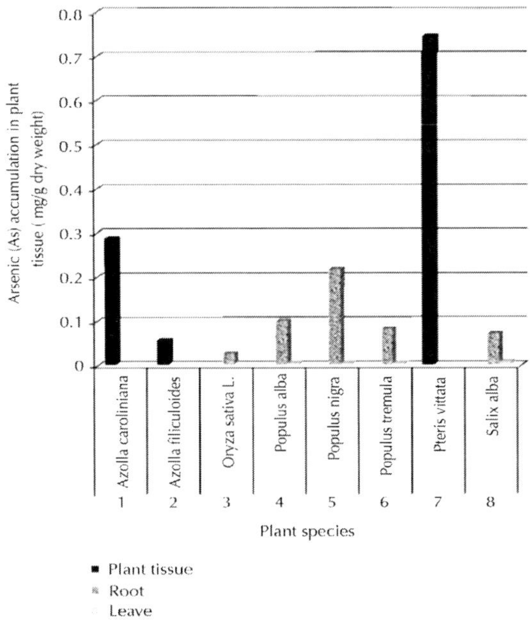

Figure 4: As accumulation in plant tissue.

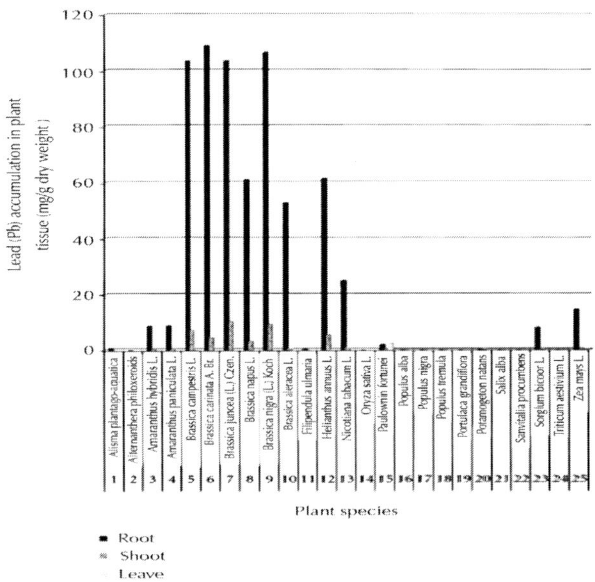

Figure 5: Pb accumulation in plant tissue.

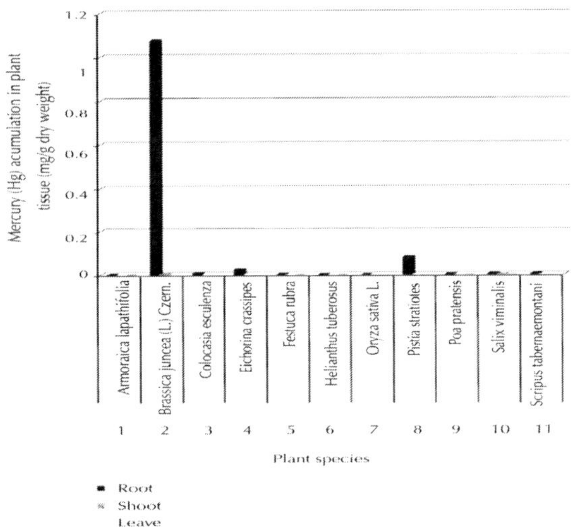

Figure 6: Hg accumulation in plant tissue.

According to Figure 4, the highest accumulation of As in plant tissue (the researchers have not detailed which part it is, but it might be the whole plant) occurs in Pteris vittata L. species. It can reach more than 0.7 mg As/g dry weight of plant. In plant root, the highest accumulation of As is in Populus nigra, which can reach more than 0.2 mg As/g dry weight of plant root.

As can be seen in Figure 5, several plants could accumulate Pb in their tissue of more than 50 mg/g dry weight of plant. Among those species are species of Brassica campestris L, Brassica carinata A. Br., Brassica juncea (L.)Czern. and Brassica nigra (L.) Koch that could accumulate more than 100 mg Pb/g dry weight.

Figure 6 shows that accumulated Hg in Brassica juncea L. Czern. is much higher than in other species of plants. It could reach more than 1 mg Hg/g dry weight of plant, while the other plants only accumulate less than 0.2 mg Hg/g dry weight.

ADVANTAGES OF PHYTOREMEDIATION

Phytoremediation techniques may also be more publicly acceptable, aesthetically pleasing, and less disruptive than the current techniques of physical and chemical process [38]. Advantages of this technology are its effectiveness in contaminant reduction, low-cost, being applicable for wide range of contaminants, and in overall it is an environmental friendly method. Figure 7 simplifies some advantages of phytoremediation technology.

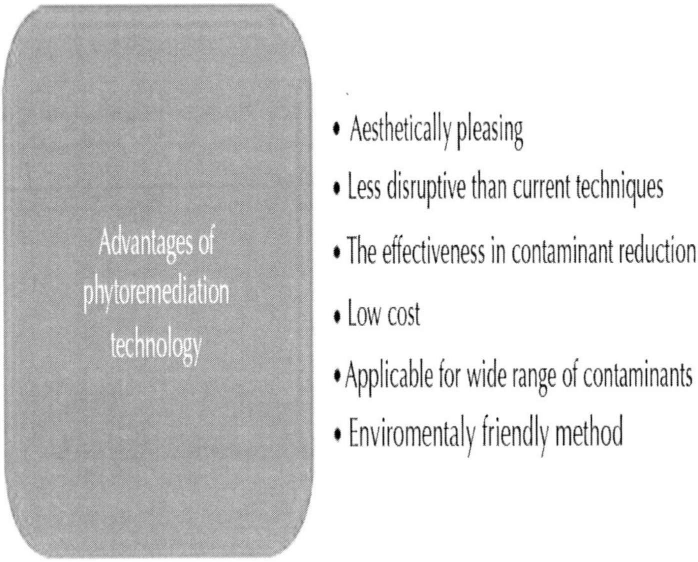

Figure 7: Advantages of phytoremediation technology.

The major advantages of the heavy metal adsorption technology by biomass are its effectiveness in reducing the concentration of heavy metal ions to very low levels and the use of inexpensive biosorbent materials [2]. Phytoremediation as possibly the cleanest and cheapest technology can be employed in the remediation of selected hazardous sites [29]. Phytoremediation encompasses a number of different methods that can lead to contaminant degradation [24].

Phytoremediation is a low-cost option and inexpensive approach for remediating environmental media, particularly suited to large sites that have relatively low levels of contamination [34]. This technology has been receiving attention lately as an innovative, cost-effective alternative to the more established treatment methods used at hazardous waste sites [29]. Phytoremediation potentially offers unique, low cost solutions to many currently problems of soil contamination [32, 75].

It is inexpensive (60–80% or even less costly) than conventional physicochemical methods, since it does not require expensive equipment or highly specialized personnel. It is cost-effective for large volumes of water having low concentrations of contaminants and for large areas having low to moderately contaminated surface soils [46].

It is applicable to a wide range of toxic metals and radionuclides [32] and also useful for treating a broad range of environmental contaminants, including organic and inorganic contaminants [46].

Phytoremediation is regarded as a new approach for the cleanup of contaminated soils, water, and ambient air [34]. Phytoremediation research can also contribute to the improvement of poor soils such as those with high aluminum or salt levels [75]. It is applicable to a range of toxic metals and radionuclides, minimal environmental disturbance, elimination of secondary air or water-borne wastes, and public acceptance [32]. Phytoextraction is considered as an environmentaly friendly method to remove metals from contaminated soils in situ. This method can be used in much larger-scale clean-up operations and has been applied for other heavy metals [76]. It is an esthetically pleasing, solar-energy-driven cleanup technology and there is minimal environmental disruption and in situ treatment preserves topsoil. In Situ applications decrease the amount of soil disturbance compared to conventional methods. It can be performed with minimal environmental disturbance with topsoil left in a usable condition and may be reclaimed for agricultural use. The organic pollutants may be degraded to CO_2 and H_2O, removing environmental toxicity [46]. Phytoremediation can be an alternative to the much harsher remediation technologies of incineration, thermal vaporization, solvent washing, or other soil washing techniques, which essentially destroy the biological component of the soil and can drastically alter its chemical and physical characteristics as well as creating a relatively nonviable solid waste. Phytoremediation actually benefits the soil, leaving an improved, functional soil ecosystem at costs estimated at approximately one-tenth of those currently adopted technologies [3]. It is the most ecological cleanup technology for contaminated soils and is also known as a green technology.

Another advantage of phytoremediation is the generation of a recyclable metal-rich plant residue [32]. Phytoremediation could be a viable option to decontaminate heavy-metal-polluted soils, particularly when the biomass produced during the phytoremediation process could be economically valorized in the form of bioenergy. The use of metal-accumulating bioenergy crops might be suitable for this purpose. If soils, contaminated with heavy metals, are phytoremediated with oil crops, biodiesel production from the resulting plant oil could be a viable option to generate bioenergy [34]. In large-scale applications, the potential energy stored can be utilized to generate thermal energy [46]. The success of the phytoextraction technique depends upon the identification of suitable plant species that can hyperaccumulate heavy metals and produce large amounts of biomass using established crop production and management practices [24].

LIMITATIONS OF PHYTOREMEDIATION TECHNOLOGY

On the other hand, there are certain limitations to phytoremediation system (Figure 8). Among them are being time-consuming method, the amount of produced biomass, the root depth, soil chemistry and the level of contamination, the age of plant, the contaminant concentration, the impacts of contaminated vegetation, and climatic condition.

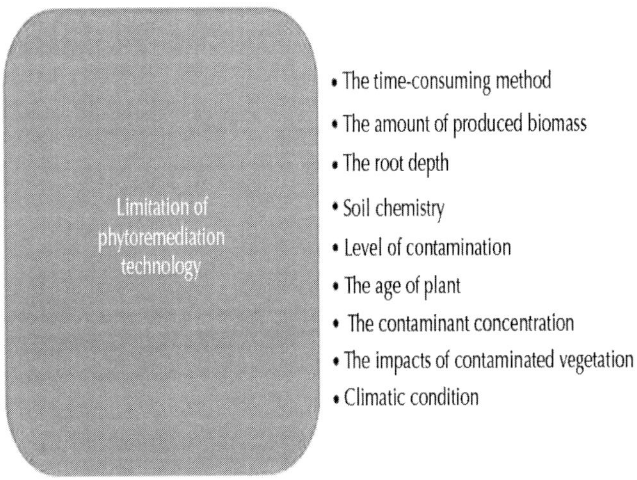

Figure 8: The limitation of phytoremediation technology.

Phytoremediation can be a time-consuming process, and it may take at least several growing seasons to clean up a site. The intermediates formed from those organic and inorganic contaminants may be cytotoxic to plants [46]. Phytoremediation is also limited by the growth rate of the plants. More time may be required to phytoremediate a site as compared with other more traditional cleanup technologies. Excavation and disposal or incineration takes weeks to months to accomplish, while phytoextraction or degradation may need several years. Therefore, for sites that pose acute risks for human and other ecological receptors, phytoremediation may not be the remediation technique of choice [29, 46]. Phytoremediation might be best suited for remote areas where human contact is limited or where soil contamination does not require an immediate response [38].

Under the best climatic conditions, with irrigation and fertilization, total biomass productivities can approach 100 t/ha/y. One of the unresolved issues is the tradeoff between toxic element accumulation and productivity. In practice, a maximum harvestable biomass yield of 10 to 20 t/ha/y would be likely, particularly for heavy metal accumulating plants. These values for productivity of biomass and heavy metal content would limit annual toxic element removal capacity between about 10 and 400 kg/ha/y, depending on the pollutant, plant species, climatic and other factors. For a target soil depth of 30 cm (4,000 t/ha), this amounts to an annual reduction from 2.5 to 100 ppm in soil toxic element levels. This is often an acceptable rate of contaminant removal, allowing site remediation over a few years to a couple of decades, particularly where the concentration of the contaminant can be lowered sufficiently to meet regulatory criteria. These values for productivity of biomass and heavy metal content would limit annual toxic element removal capacity between 10 and 400 kg/ha/y, depending on the pollutant, plant species, climatic and other factors [37].

The success of phytoremediation may be limited by factors such as growing time, climate, root depth, soil chemistry, and level of contamination [38]. Root contact is a primary limitation on phytoremediation applicability. Remediation with plants requires that contaminants be in contact with the root zone of the plants. Either the plants must be able to extend roots to the contaminants, or the contaminated media must be moved to be within range of the plants [29]. Restricted to sites with shallow contamination within rooting zone of remediative plants, ground surface at the site may have to be modified to prevent flooding or erosion [46].

Age greatly affects the physiological activity of a plant, especially its roots. Generally, roots of a young plant display greater ability to absorb ions than do those of an old plant when they are similar in size. It is important to use healthy young plants for more efficient plant removal. However, this does not rule out the use of larger older plants whose larger size may compensate for their lower physiological activity as compared to smaller younger plants [45].

High concentrations of contaminants may inhibit plant growth and, thus, may limit application on some sites or some parts of sites. This phytotoxicity could lead to a remedial approach in which high-concentration waste is handled with expensive ex situ techniques that quickly reduce acute risk, while in situ phytoremediation is used over a longer period of time to clean the high volumes of lower contaminant concentrations [29]. A major limitation in the phytoremediation of toxic elements is the maximal level that can be accumulated by plants. Plants with the highest levels of toxic metal contents, known as "hyperaccumulators", generally exhibit, on a dry weight basis, from about 2000 ppm (0.2%) for more toxic elements (Cd, Pb) to above 2% for the less toxic ones (Zn, Ni, Cu) [75]. Restricted to sites with low contaminant concentrations, the treatment is generally limited to soils at one meter from the surface and groundwater within a few meters of the surface with soil amendments may be required [46].

Some ecological exposure may occur whenever plants are used to interact with contaminants from the soil. The fate of the metals in the biomass is a concern. Although some forms of phytoremediation involve accumulation of metals and require handling of plant material embedded with metals, most plants do not accumulate significant levels of organic contaminants. While metal-accumulating plants will need to be harvested and either recycled or disposed of in compliance with applicable regulations, most phytoremediative plants do not require further treatment or disposal [29]. Harvested plant biomass from phytoextraction may be classified as a hazardous waste, hence, disposal should be proper. Consumption of contaminated plant biomass is a cause of concern; contaminants may still enter the food chain through animals/insects that eat plant material containing contaminants [46].

Climatic or hydrologic conditions may restrict the rate of growth of plants that can be utilized. Introduction of nonnative species may affect biodiversity [46].\

CONCLUSIONS

Heavy metals uptake, by plants using phytoremediation technology, seems to be a prosperous way to remediate heavy-metals-contaminated environment. It has some advantages compared with other commonly used conventional technologies. Several factors must be considered in order to accomplish a high performance of remediation result. The most important factor is a suitable plant species which can be used to uptake the contaminant. Even the phytoremediation technique seems to be one of the best alternative, it also has some limitations. Prolong research needs to be conducted to minimize this limitation in order to apply this technique effectively.

REFERENCES

1. A. Gaur and A. Adholeya, "Prospects of arbuscular mycorrhizal fungi in phytoremediation of heavy metal contaminated soils," Current Science, vol. 86, no. 4, pp. 528–534, 2004.
2. R. Rakhshaee, M. Giahi, and A. Pourahmad, "Studying effect of cell wall's carboxyl-carboxylate ratio change of Lemna minor to remove heavy metals from aqueous solution," Journal of Hazardous Materials, vol. 163, no. 1, pp. 165–173, 2009.
3. R. R. Hinchman, M. C. Negri, and E. G. Gatliff, "Phytoremediation: using green plants to clean up contaminated soil, groundwater, and wastewater," Argonne National Laboratory Hinchman, Applied Natural Sciences, Inc, 1995, http://www.treemediation.com/Technical/Phytoremediation_1998.pdf.
4. I. Shtangeeva, J. V.-P. Laiho, H. Kahelin, and G. R. Gobran, "Phytoremediation of metal-contaminated soils. Symposia Papers Presented Before the Division of Environmental Chemistry," American Chemical Society, Anaheim, Calif, USA, 2004,http://ersdprojects.science.doe.gov/workshop_pdfs/california_2004/p050.pdf.
5. K. Cho-Ruk, J. Kurukote, P. Supprung, and S. Vetayasuporn, "Perennial plants in the phytoremediation of lead-contaminated soils," Biotechnology, vol. 5, no. 1, pp. 1–4, 2006.

6. M. M. Lasat, "Phytoextraction of metals from contaminated soil: a review of plant/soil/metal interaction and assessment of pertinent agronomic issues," Journal of Hazardous Substance Research, vol. 2, no. 5, pp. 1–25, 2000.
7. E. Pehlivan, A. M. Özkan, S. Dinç, and S. Parlayici, "Adsorption of Cu_{2+} and Pb_{2+} ion on dolomite powder," Journal of Hazardous Materials, vol. 167, no. 1–3, pp. 1044–1049, 2009.
8. S. Roy, S. Labelle, P. Mehta et al., "Phytoremediation of heavy metal and PAH-contaminated brownfield sites," Plant and Soil, vol. 272, no. 1-2, pp. 277–290, 2005.
9. D. Mohan and C. U. Pittman Jr., "Arsenic removal from water/wastewater using adsorbents—a critical review," Journal of Hazardous Materials, vol. 142, no. 1-2, pp. 1–53, 2007.
10. National Ground Water Association, Copyright 2001. Arsenic. What you need to knowhttp://www.ngwa.org/ASSETS/A0DD107452D74B33AE9D5114EE6647ED/Arsenic.pdf.
11. U.S. Department of Health and Human Services, Public Health Service Agency for Toxic Substances and Disease Registry. Division of Toxicology and Environmental Medicine. Arsenic. 2005,http://www.baltimorehealth.org/info/ATSDR%20fact%20sheet.pdf.
12. H. Hasegawa, M. A. Rahman, T. Matsuda, T. Kitahara, T. Maki, and K. Ueda, "Effect of eutrophication on the distribution of arsenic species in eutrophic and mesotrophic lakes," Science of the Total Environment, vol. 407, no. 4, pp. 1418–1425, 2009. · ·
13. WHO Regional Office for Europe, Air Quality Guidelines, chapter 6.1, Arsenic, Copenhagen, Denmark, 2nd edition, 2000, http://www.euro.who.int/document/aiq/6_1_arsenic.pdf.
14. P. Chutia, S. Kato, T. Kojima, and S. Satokawa, "Arsenic adsorption from aqueous solution on synthetic zeolites," Journal of Hazardous Materials, vol. 162, no. 1, pp. 440–447, 2009.
15. H. A. Andrianisa, A. Ito, A. Sasaki, J. Aizawa, and T. Umita, "Biotransformation of arsenic species by activated sludge and removal of bio-oxidised arsenate from wastewater by coagulation with ferric chloride," Water Research, vol. 42, no. 19, pp. 4809–4817, 2008.

16. R. J. Ampiah-Bonney, J. F. Tyson, and G. R. Lanza, "Phytoextraction of arsenic from soil by Leersia oryzoides," International Journal of Phytoremediation, vol. 9, no. 1, pp. 31–40, 2007.
17. M. Vaclavikova, G. P. Gallios, S. Hredzak, and S. Jakabsky, "Removal of arsenic from water streams: an overview of available techniques," Clean Technologies and Environmental Policy, vol. 10, no. 1, pp. 89–95, 2008.
18. A. M. Yusof and N. A. N. N. Malek, "Removal of Cr(VI) and As(V) from aqueous solutions by HDTMA-modified zeolite Y," Journal of Hazardous Materials, vol. 162, no. 2-3, pp. 1019–1024, 2009.
19. WHO Regional Office for Europe, Air Quality Guidelines, chapter 6.7, Lead, Copenhagen, Denmark, 2nd edition, 2001, http://www.euro.who.int/document/aiq/6_7lead.pdf.
20. J. H. Traunfeld and D. L. Clement, "Lead in Garden Soils. Home and Garden," Maryland Cooperative Extention, University of Maryland, 2001, http://www.hgic.umd.edu/_media/documents/hg18.pdf.
21. European Commission DG ENV. E3. Heavy Metals in Waste, Final Report Project ENV.E.3/ETU/2000/0058, 2002http://ec.europa.eu/environment/waste/studies/pdf/heavy_metalsreport.pdf.
22. J. F. Musselman and QEP, "Sources of Mercury in Wastewater, Pretreatment corner," http://www.cet-inc.com/cmsdocuments//7%20-%20Sources%20of%20Mercury%20in%20Wastewater%20(0204).pdf.
23. T. C. Chang, S. J. You, B. S. Yu, C. M. Chen, and Y. C. Chiu, "Treating high-mercury-containing lamps using full-scale thermal desorption technology," Journal of Hazardous Materials, vol. 162, no. 2-3, pp. 967–972, 2009.
24. L. Rodriguez, F. J. Lopez-Bellido, A. Carnicer, F. Recreo, A. Tallos, and J. M. Monteagudo, "Mercury recovery from soils by phytoremediation," in Book of Environmental Chemistry, pp. 197–204, Springer, Berlin, Germany, 2005.
25. I. Wagner-Döbler, "Microbiological treatment of mercury-loaded waste water—the biological mercury-decontamination-system. German Research Centre for Biotechnology,"http://www.gbf.de/mercury_remediation1/pdf-documents/Information%20leaflet.PDF.

26. A. Resaee, J. Derayat, S. B. Mortazavi, Y. Yamini, and M. T. Jafarzadeh, "Removal of Mercury from chlor-alkali industry wastewater using Acetobacter xylinum cellulose," American Journal of Environmental Sciences, vol. 1, no. 2, pp. 102–105, 2005.
27. A. Sas-Nowosielska, R. Galimska-Stypa, R. Kucharski, U. Zielonka, E. Małkowski, and L. Gray, "Remediation aspect of microbial changes of plant rhizosphere in mercury contaminated soil,"Environmental Monitoring and Assessment, vol. 137, no. 1–3, pp. 101–109, 2008.
28. A. Erakhrumen and A. Agbontalor, "Review Phytoremediation: an environmentally sound technology for pollution prevention, control and remediation in developing countries," Educational Research and Review, vol. 2, no. 7, pp. 151–156, 2007.
29. U. S. Environmental Protection Agency, "Introduction to Phytoremediation," National Risk Management Research Laboratory, EPA/600/R-99/107, 2000, http://www.clu-in.org/download/remed/introphyto.pdf.
30. F. N. Moreno, C. W. N. Anderson, R. B. Stewart, and B. H. Robinson, "Phytofiltration of mercury-contaminated water: volatilisation and plant-accumulation aspects," Environmental and Experimental Botany, vol. 62, no. 1, pp. 78–85, 2008.
31. M. N. V. Prasad and H. M. De Oliveira Freitas, "Metal hyperaccumulation in plants—biodiversity prospecting forphytoremediation technology," Electronic Journal of Biotechnology, vol. 6, no. 3, pp. 110–146, 2003.
32. D. Liu, W. Jiang, C. Liu, C. Xin, and W. Hou, "Uptake and accumulation of lead by roots, hypocotyls and shoots of Indian mustard [Brassica juncea (L.)]," Bioresource Technology, vol. 71, no. 3, pp. 273–277, 2000. ·
33. T. Bhattacharya, D. K. Banerjee, and B. Gopal, "Heavy metal uptake by Scirpus littoralis Schrad. from fly ash dosed and metal spiked soils," Environmental Monitoring and Assessment, vol. 121, no. 1-3, pp. 363–380, 2006.
34. L. Van Ginneken, E. Meers, R. Guisson et al., "Phytoremediation for heavy metal-contaminated soils combined with bioenergy production," Journal of Environmental Engineering and Landscape Management, vol. 15, no. 4, pp. 227–236, 2007.

35. Interstate Technology and Regulatory Council, Phytotechnology Technical and Regulatory. Guidance and Decision Trees, 2009, http://www.itrcweb.org/guidancedocument.asp?TID=63.
36. R. K. Sinha, S. Herat, and P. K. Tandon, "14 phytoremediation: role of plants in contaminated site management," in Book of Environmental Bioremediation Technologies, pp. 315–330, Springer, Berlin, Germany, 2004.
37. U. S. Department of Energy, "Plume Focus Area, December. Mechanisms of plant uptake, translocation, and storage of toxic elements. Summary Report of a workshop on phytoremediation research needs," 1994,http://www.osti.gov/bridge/purl.cover.jsp;jsessionid=D72C8DD9003DCF51984EE254A6ED8BCB?purl=/10109412-BckU4U/webviewable/.
38. A. L. Salido, K. L. Hasty, J. M. Lim, and D. J. Butcher, "Phytoremediation of arsenic and lead in contaminated soil using Chinese Brake ferns (Pteris vittata) and Indian mustard (Brassica juncea),"International Journal of Phytoremediation, vol. 5, no. 2, pp. 89–103, 2003.
39. L. Erdei, G. Mezôsi, I. Mécs, I. Vass, F. Fôglein, and L. Bulik, "Phytoremediation as a program for decontamination of heavy-metal polluted environment," in Proceedings of the 8th Hungarian Congress on Plant Physiology and the 6th Hungarian Conference on Photosynthesis, 2005.
40. L. Erdei, G. Mezôsi, I. Mécs, I. Vass, F. Fôglein, and L. Bulik, "Phytoremediation as a program for decontamination of heavy-metal polluted environment," Acta Biologica Szegediensis, vol. 49, no. 1-2, pp. 75–76, 2005.
41. U. S. Environmental Protection Agency, Use of Field-Scale Phytotechnology, for Chlorinated Solvents, Metals, Explosives, and Propellants, and Pesticides Phytotechnology Mechanisms. Solid Waste and Emergency Response (5102G), EPA 542-R-05-002, 2005, http://www.clu-in.org/download/remed/542-r-05-002.pdf.
42. V. M. Ibeanusi, "Denise Antonia Grab In collaboration with Larry, Jensen Stephen Ostrodka—Environmental Protection Agency. Radionuclide Biological Remediation Resource Guide, U. S. Environmental Protection Agency," 2004, http://www.clu-in.org/download/remed/905b04001.pdf.

43. N. Merkl, R. Schultze-Kraft, and C. Infante, "Phytoremediation in the tropics—influence of heavy crude oil on root morphological characteristics of graminoids," Environmental Pollution, vol. 138, no. 1, pp. 86–91, 2005.
44. J. G. Burken and J. L. Schnoor, "Phytoremediation: plant uptake of atrazine and role of root exudates,"Journal of Environmental Engineering, vol. 122, no. 11, pp. 958–963, 1996.
45. S. Tu, L. Q. Ma, A. O. Fayiga, and E. J. Zillioux, "Phytoremediation of arsenic-contaminated groundwater by the arsenic hyperaccumulating fern Pteris vittata L," International Journal of Phytoremediation, vol. 6, no. 1, pp. 35–47, 2004.
46. W. J. S. Mwegoha, "The use of phytoremediation technology for abatement soil and groundwater pollution in Tanzania: opportunities and challenges," Journal of Sustainable Development in Africa, vol. 10, no. 1, pp. 140–156, 2008.
47. A. Fritioff and M. Greger, "Aquatic and Terrestrial Plant Species with Potential to Remove Heavy Metals from Stormwater," International Journal of Phytoremediation, vol. 5, no. 3, pp. 211–224, 2003.
48. P. Seuntjens, B. Nowack, and R. Schulin, "Root-zone modeling of heavy metal uptake and leaching in the presence of organic ligands," Plant and Soil, vol. 265, no. 1-2, pp. 61–73, 2004.
49. R. Chandra, R. N. Bharagava, S. Yadav, and D. Mohan, "Accumulation and distribution of toxic metals in wheat (Triticum aestivum L.) and Indian mustard (Brassica campestris L.) irrigated with distillery and tannery effluents," Journal of Hazardous Materials, vol. 162, no. 2-3, pp. 1514–1521, 2009.
50. M. Gupta, P. Sharma, N. B. Sarin, and A. K. Sinha, "Differential response of arsenic stress in two varieties of Brassica juncea L," Chemosphere, vol. 74, no. 9, pp. 1201–1208, 2009.
51. D. Hammer, A. Kayser, and C. Keller, "Phytoextraction of Cd and Zn with Salix viminalis in field trials,"Soil Use and Management, vol. 19, no. 3, pp. 187–192, 2003.
52. R. E. Hamon, P. E. Holm, S. E. Lorenz, S. P. McGrath, and T. H. Christensen, "Metal uptake by plants from sludge-amended soils: caution is required in the plateau interpretation," Plant and Soil, vol. 216, no. 1-2, pp. 53–64, 1999.

53. M. S. Liphadzi, M. B. Kirkham, K. R. Mankin, and G. M. Paulsen, "EDTA-assisted heavy-metal uptake by poplar and sunflower grown at a long-term sewage-sludge farm," Plant and Soil, vol. 257, no. 1, pp. 171–182, 2003.
54. A. Murányi and L. Ködöböcz, "Heavy metal uptake by plants in different phytoremediation treatments," in Proceedings of the 7th Alps-Adria Scientific Workshop, Stara Lesna, Slovakia, 2008,http://www.mokkka.hu/publications/0387.117_MOKKA_AM_LK.pdf.
55. I. D. Pulford, D. Riddell-Black, and C. Stewart, "Heavy metal uptake by willow clones from sewage sludge-treated soil: the potential for phytoremediation," International Journal of Phytoremediation, vol. 4, no. 1, pp. 59–72, 2002.
56. J. Rydlová and M. Vosátka, "Effect of Glomus intraradices isolated from PB-contaminated soil on PB uptake by Agrostis capillaris is changed by its cultivation in a metal-free substrate," Folia Geobotanica, vol. 38, no. 2, pp. 155–165, 2003.
57. L. Sebastiani, F. Scebba, and R. Tognetti, "Heavy metal accumulation and growth responses in poplar clones Eridano (Populus deltoides x maximowiczii) and I-214 (P. x euramericana) exposed to industrial waste," Environmental and Experimental Botany, vol. 52, no. 1, pp. 79–88, 2004.
58. S. Sharma, "Study on impact of heavy metal accumulation in Brachythecium populeum (Hedw.) B.S.G,"Ecological Indicators, vol. 9, no. 4, pp. 807–811, 2009.
59. T. Vamerali, M. Bandiera, L. Coletto, F. Zanetti, N. M. Dickinson, and G. Mosca, "Phytoremediation trials on metal- and arsenic-contaminated pyrite wastes (Torviscosa, Italy)," Environmental Pollution, vol. 157, no. 3, pp. 887–894, 2009.
60. T. Vamerali, M. Bandiera, L. Coletto, F. Zanetti, N. M. Dickinson, and G. Mosca, "Phytoremediation trials on metal- and arsenic-contaminated pyrite wastes (Torviscosa, Italy)," Environmental Pollution, vol. 157, no. 3, pp. 887–894, 2009.
61. B. Vandecasteele, E. Meers, P. Vervaeke, B. D. Vos, P. Quataert, and F. M. G. Tack, "Growth and trace metal accumulation of two Salix clones on sediment-derived soils with increasing contamination levels,"Chemosphere, vol. 58, no. 8, pp. 995–1002, 2005.

62. P. Vervaeke, F. M. G. Tack, N. Lust, and M. Verloo, "Short- and longer-term effects of the willow root system on metal extractability in contaminated dredged sediment," Journal of Environmental Quality, vol. 33, no. 3, pp. 976–983, 2004.
63. M. Vyslouzilova, P. Tlustos, J. Szakova, and D. Pavlikova, "As, Cd, Pb and Zn uptake by Salix spp.clones grown in soil enrich by high load of this elements," Plant Soil Environment, vol. 49, no. 5, pp. 191–196, 2003.
64. H. B. Wang, Z. H. Ye, W. S. Shu, W. C. Li, M. H. Wong, and C. Y. Lan, "Arsenic uptake and accumulation in fern species growing at arsenic-contaminated sites of Southern China: field surveys,"International Journal of Phytoremediation, vol. 8, no. 1, pp. 1–11, 2006.
65. J. Wang, C. B. Zhang, and Z. X. Jin, "The distribution and phytoavailability of heavy metal fractions in rhizosphere soils of Paulowniu fortunei (seem) Hems near a Pb/Zn smelter in Guangdong, PR China,"Geoderma, vol. 148, no. 3-4, pp. 299–306, 2009.
66. W.-X. Liu, L.-F. Shen, J.-W. Liu, Y.-W. Wang, and S.-R. Li, "Uptake of toxic heavy metals by rice (Oryza sativa L.) cultivated in the agricultural soil near Zhengzhou City, People›s Republic of China," Bulletin of Environmental Contamination and Toxicology, vol. 79, no. 2, pp. 209–213, 2007.
67. M. Ebrahimpour and I. Mushrifah, "Heavy metal concentrations (Cd, Cu and Pb) in five aquatic plant species in Tasik Chini, Malaysia," Environmental Geology, vol. 54, no. 4, pp. 689–698, 2008.
68. R. Feng, C. Wei, S. Tu, and X. Sun, "Interactive effects of selenium and arsenic on their uptake by Pteris vittata L. under hydroponic conditions," Environmental and Experimental Botany, vol. 65, no. 2-3, pp. 363–368, 2009.
69. M. A. Rahman, H. Hasegawa, K. Ueda, T. Maki, and M. M. Rahman, "Arsenic uptake by aquatic macrophyte Spirodela polyrhiza L.: interactions with phosphate and iron," Journal of Hazardous Materials, vol. 160, no. 2-3, pp. 356–361, 2008
70. E. Rodríguez, J. R. Peralta-Videa, M. Israr et al., "Effect of mercury and gold on growth, nutrient uptake, and anatomical changes in

Chilopsis linearis," Environmental and Experimental Botany, vol. 65, no. 2-3, pp. 253–262, 2009. ·

71. K. Skinner, N. Wright, and E. Porter-Goff, "Mercury uptake and accumulation by four species of aquatic plants," Environmental Pollution, vol. 145, no. 1, pp. 234–237, 2007.
72. M. Srivastava, L. Q. Ma, B. Rathinasabapathi, and P. Srivastava, "Effects of selenium on arsenic uptake in arsenic hyperaccumulator Pteris vittata L," Bioresource Technology, vol. 100, no. 3, pp. 1115–1121, 2009. · ·
73. C. Watson, I. D. Pulford, and D. Riddell-Black, "Screening of willow species for resistance to heavy metals: comparison of performance in a hydroponics system and field trials," International Journal of Phytoremediation, vol. 5, no. 4, pp. 351–365, 2003.
74. X. Zhang, A. J. Lin, F. J. Zhao, G. Z. Xu, G. L. Duan, and Y. G. Zhu, "Arsenic accumulation by the aquatic fern Azolla: comparison of arsenate uptake, speciation and efflux by A. caroliniana and A. filiculoides," Environmental Pollution, vol. 156, no. 3, pp. 1149–1155, 2008.
75. US Department of Energy, "Plume Focus Area. Summary Report of a Workshop on Phytoremediation Research Needs, Office of Technology Development, Office of Environmental Management and Division of Energy Biosciences, Office of Basic Energy Sciences, Office of Energy Research," 1994,http://www.law.csuohio.edu/lawlibrary-oldsite/info_services/acquisitions/acq-0104.html.
76. Y. Wang and M. Greger, "Use of iodide to enhance the phytoextraction of mercury-contaminated soil,"Science of the Total Environment, vol. 368, no. 1, pp. 30–39, 2006.

Chapter 4

Spatial Patterns and Temporal Trends of Heavy Metal Concentrations in Moss and Surface Soil Specimens Collected in Norway between 1990 and 2010

Michaela Meyer[1], Roland Pesch[1], Winfried Schröder[1], Eiliv Steinnes[2], and Hilde Thelle Uggerud[3]

[1]Landscape Ecology, University of Vechta, Vechta, 49377, Germany

[2]Department of Chemistry, Norwegian University of Science and Technology, Trondheim, NO-7491, Norway

[3]Norwegian Institute for Air Research (NILU), Kjeller, 2027, Norway

ABSTRACT

Background

The aim of this investigation was to inquire whether the spatial patterns and temporal trends of heavy metal concentrations in moss and soil specimen monitored in two spatial dense networks covering Norway are correlated. To this end, data about concentrations of cadmium, mercury and lead in moss and soil specimens collected were compiled. The data were derived from moss surveys conducted in 1990, 1995, 2000, 2005 and 2010, as well as from soil monitoring campaigns in 1995 and 2005.

Results

The data sets from both moss and soil surveys indicate a decrease of heavy metal concentrations in moss and soil specimen. However, in case of moss samples, the decrease is by far more pronounced and statistically significant. The heavy metal concentrations in moss and soil are correlated significantly with high positive coefficients for Pb, medium for Cd and moderate for Hg. From a set of potentially influencing boundary conditions, the modelled atmospheric deposition showed the highest correlation with the heavy metal concentrations in moss and soil. The spatial patterns of Cd and Pb concentration in moss and soil specimens 1995 and 2005 are similar. Thereby, the spatial differentiation of concentrations in moss is higher than that in soil, while the opposite holds true for the Hg concentration.

Conclusions

Even if the metal concentrations in moss and soil are statistically correlated, they should not be replaced by each other but should be used as complementary monitoring systems.

BACKGROUND

Environmental exposure to pollutants such as heavy metals (HM) can impact structures and functions being relevant for ecosystem integrity and human health [1]-[6]. To avoid such impacts due to atmospheric HM deposition, the Protocol on Heavy Metals of the Convention on Long-Range Transboundary Air Pollution, adopted in 1998, targets three particularly harmful metals: cadmium (Cd), lead (Pb) and mercury (Hg). Accordingly, Parties have to reduce the respective emissions from industrial sources, combustion and waste incineration below the respective levels in 1990 or an alternative year between 1985 and 1995. As calculated by [7], the European ecosystem area at risk in 2000 of emissions of cadmium, mercury and lead was estimated to amount for about 0.34%, 77% and 42%, respectively. In 2020, these areas are expected to reduce to 0.02%, 74% and 19%, respectively. Thus, the atmospheric deposition of mercury and lead in particular puts large areas of European ecosystems at risk of adverse effects both in 2000 and 2020. The effect-based scenario analysis of emissions of chromium, nickel, copper, zinc, arsenic and selenium indicates that a relatively small ecosystem area in Europe is subject to risk of adverse effects of these metals in 2000. The atmospheric deposition of copper, zinc and selenium is expected to cause risk of adverse effects of 1% at maximum of the European ecosystem area in 2020. However, modelled atmospheric deposition of copper, zinc and selenium are significantly underestimated. Nevertheless, none of the metals other than Cd, Hg and Pb achieve high enough concentrations due to atmospheric deposition to cause adverse effects on ecosystems and human health. Thus, the focus on Cd, Hg and Pb as priority metals is regarded as justified [7]. Accordingly, the paper at hand focuses on these three metals.

To validate modelled atmospheric deposition values and risk estimations mentioned above, measurements of atmospheric deposition and complementary estimates from element concentration in mosses are well-established methods [2],[5],[8]-[11]. This is due to specific properties: Since soil-inhabiting ectohydric moss species do not have real roots and a weakly developed cuticle, grow slow and show minimal morphological changes during lifetime, element concentrations measured in moss integrates fairly well dry, occult and wet atmospheric deposition. Moss can accumulate large amounts of atmospheric HM deposition due to efficient element capture (high surface-to-volume ratio) and retention as well as resistance to toxic elements [8],[10],[12]-[16]. Elements are deposited on moss in aqueous solution, in gaseous form or attached to particles. The bioaccessible HM fraction is the portion of environmentally available metal that interacts at the moss contact surface and is potentially available for adsorption and absorption by the moss. Once adsorbed, HM can be absorbed into and across the biological surface and distributed throughout the organism. The net accumulation of HM in moss that results from exposure to environmentally available HM is called bioaccumulation, including both metal adsorbed to surfaces or absorbed by the organism, expressed on a weight (dry or wet)-adjusted basis [17]. HM accumulate in mosses as layers of particles or entrapment on the surface of the cells, through incorporation into the outer walls of the cells (ion exchange processes), and metabolically controlled passage into the cells. The sorbed metal may be located either in the extracellular region outside of the cytoplasm or bound to the cell wall, and ions have a direct access to the cell wall since mosses do not discriminate HM ions [8].

In Norway, the monitoring of atmospheric HM deposition has been set up since the 1980s and is continued until today at three stations [18]. Due to the limited spatial coverage of the deposition monitoring network in Norway, the spatial and temporal mapping of HM deposition was carried out by means of sampling and analysing moss. First established and implemented in 1970, by means of moss [19] could show a great geographical variety of some HM within Norway, with considerable higher values in the southern part. Therefore, an extensive moss survey with 500 sampling sites distributed all over Norway was carried out in 1977 [20]. In 1985, it was decided to reproduce the moss survey every 5 years with additional surveys in other Nordic countries. Since 1990, further nationwide surveys in at least 21 European countries are carried out every 5 years with an increasing number of HM elements due to improvements in analytical procedures [18],[21]-[24]. Following [25], mapping atmospheric HM deposition from the atmosphere to terrestrial ecosystems by means of mosses can provide a complementary and time-integrated method with a high spatial and temporal resolution.

From the beginning, Norway has incorporated a great number of HM elements compared to concurrent moss surveys in other countries. This paper deals both with spatial patterns and temporal trends of concentrations of Hg, Cd and Pb in moss and soil specimen as well as with potential influencing environmental factors and aimed to investigate whether the spatial patterns and temporal trends of heavy metal concentrations in moss and soil specimen monitored in two spatial dense networks covering Norway are correlated.

RESULTS AND DISCUSSION

Temporal Trend Analyses

Table 1 depicts descriptive statistical measures for Cd, Hg and Pb in moss sampled over the period 1990 to 2010, and Table 2 shows the results of the Wilcoxon signed-rank test in terms of the calculated *p* values. Figures 1, 2 and 3 furthermore show the development of the moss concentrations as box plots. To more clearly illustrate the temporal trend for Cd, Hg and Pb outliers are blinded out of the box plots.

Table 1: Descriptive statistical measures for Cd, Hg and Pb concentrations in moss and in soil

Element	Year	Number	Median	Minimum	Maximum
Cd (µg/g) moss	1990	495	0.126	0.026	3.373
	1995	458	0.128	0.004	1.672
	2000	464	0.087	0.002	2.622
	2005	464	0.089	0.017	2.453
	2010	463	0.081	0.009	1.875
Hg (µg/g) moss	1990	151	0.068	0.014	0.817
	1995	209	0.068	0.002	0.479
	2000	464	0.052	0.022	0.208
	2005	464	0.054	0.019	0.250
	2010	460	0.064	0.025	0.445
Pb (µg/g) moss	1990	495	9.3	1.5	78.9
	1995	458	5.8	0.9	59.4
	2000	464	2.7	0.5	27.7
	2005	464	2.2	0.5	34.3
	2010	463	1.5	0.3	20.8

Cd (μg/g) soil	1995	461	0.45	0.05	6.66
	2005	462	0.40	0.05	5.72
Hg (μg/g) soil	1995	No data			
	2005	462	0.20	0.03	8.86
Pb (μg/g) soil	1995	461	34.56	7.05	798.79
	2005	462	25.75	3.80	1259.80

The descriptive statistical measures for Cd, Hg and Pb concentrations in moss are from 1990 to 2010 and in soil from 1995 and 2005.

Meyer *et al.*

Meyer *et al.* *Environmental Sciences Europe* 2014 26:27 doi:10.1186/s12302-014-0027-0

Table 2: Results of Wilcoxon signed-rank test for Cd, Hg, and Pb concentration in moss and soil

Years	Cd	Hg	Pb
Moss			
1990 to 1995	0.60	0.80	0.00
1990 to 2000	0.00	0.00	0.00
1990 to 2005	0.08	0.00	0.00
1990 to 2010	0.00	0.65	0.00
1995 to 2000	0.00	0.00	0.00
1995 to 2005	0.00	0.00	0.00
1995 to 2010	0.00	0.30	0.00
2000 to 2005	0.00	0.09	0.00
2000 to 2010	0.52	0.00	0.00
2005 to 2010	0.00	0.00	0.00
Soil			
1995 to 2005	0.21	No data	0.00

The significance level is given by p-values. The italicized values signify $p > 0.01$ (not significant).

Meyer *et al.*

Meyer *et al.* *Environmental Sciences Europe* 2014 26:27 doi:10.1186/s12302-014-0027-0

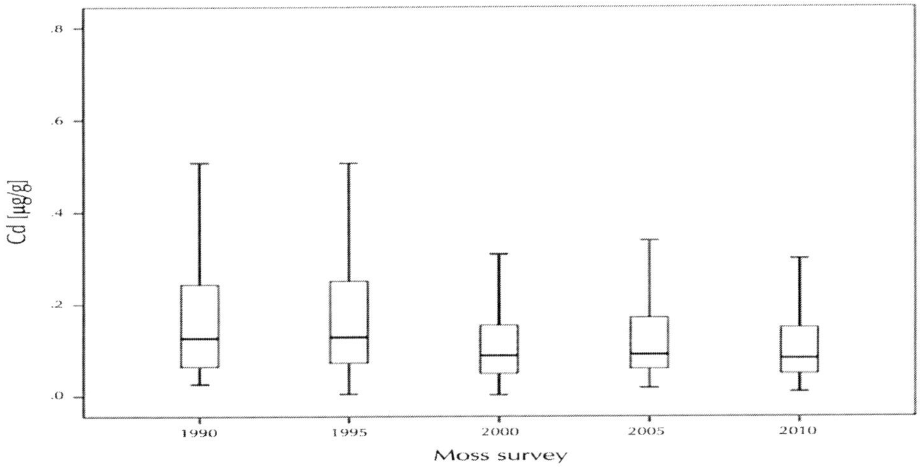

Figure 1: Box plot for Cd in moss, Norway 1990 to 2010 (no outliers depicted).

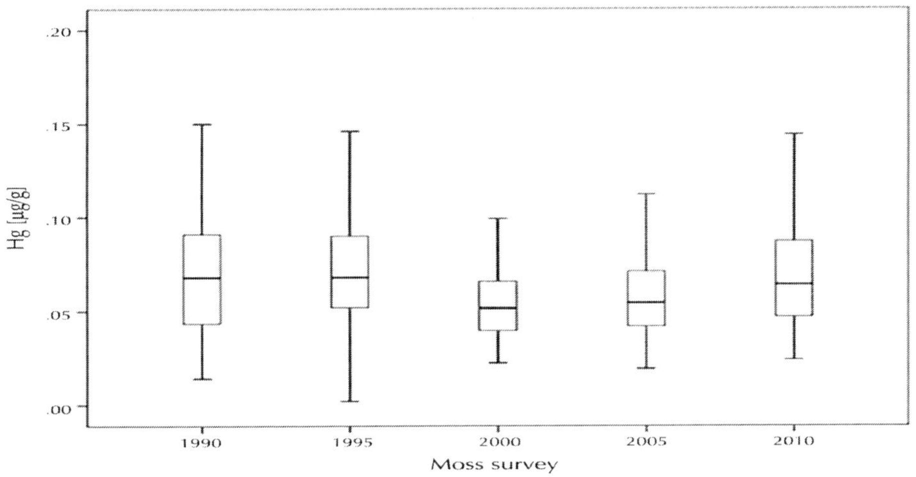

Figure 2: Box plot for Hg in moss, Norway 1990 to 2010 (no outliers depicted).

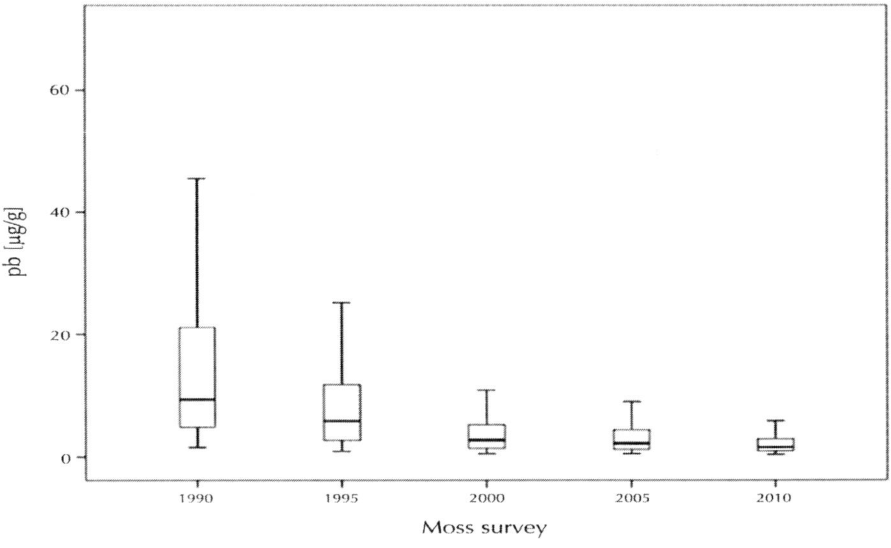

Figure 3: Box plot for Pb in moss, Norway 1990 to 2010 (no outliers depicted).

As can be seen from Table 1 and Figures 1, 2 and 3, the highest medians as well as maximum values in moss can be observed in the 1990 survey for all three elements. A clear continuous decrease from 1990 to 2010 can be observed for Pb only reaching from 9.3 µg/g in 1990 to 1.5 µg/g in 2010. Regarding Hg, in 2010, a median concentration of 0.064 µg/g was calculated. This almost reaches the same concentration of Hg in moss as in 1990 and 1995 (0.068 µg/g). The inference statistical comparison of all three elements shows significant differences between the years in all cases except for Cd between 1990 and 1995, 1990 and 2005and 2000 and 2010 as well as for Hg between 1990 and 1995, 1990 and 2010, 1995 and 2010 and 2000 and 2005 (alpha =0.01) (Table 2). As can be seen from Table 1 and Figures 4, 5 and 6, the highest median values in soil can be observed in 1995 for Cd and Pb. There were no data for Hg in 1995. In case of Pb, there was a slight but significant decrease over time reaching from 34.6 µg/g in 1995 to 25.8 µg/g in 2005 (alpha =0.01) (Table 2).

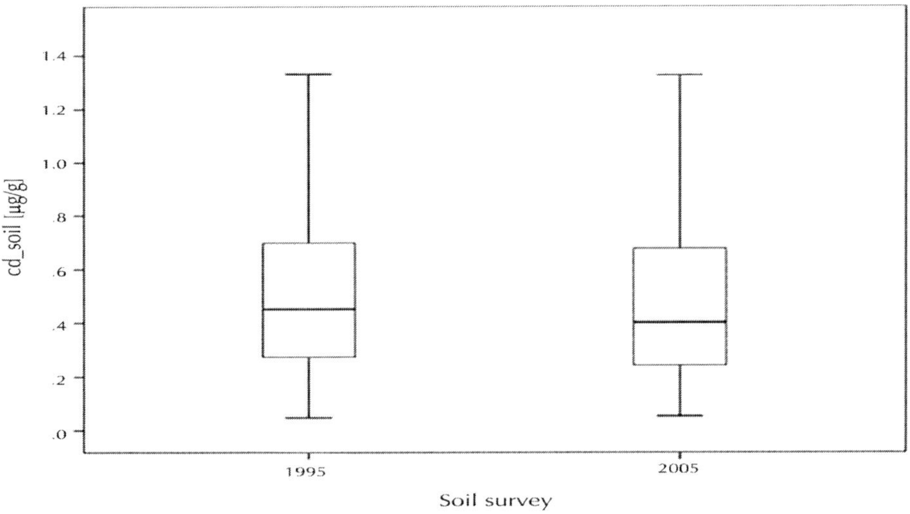

Figure 4: Box plot for Cd in soil, Norway 1995 and 2005 (no outliers depicted).

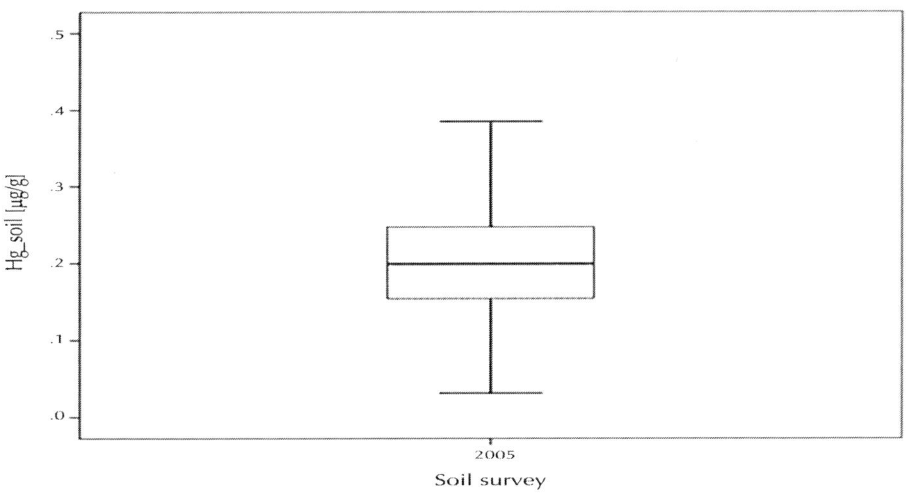

Figure 5: Box plot for Hg in soil, Norway 2005 (no outliers depicted, no data for 1995).

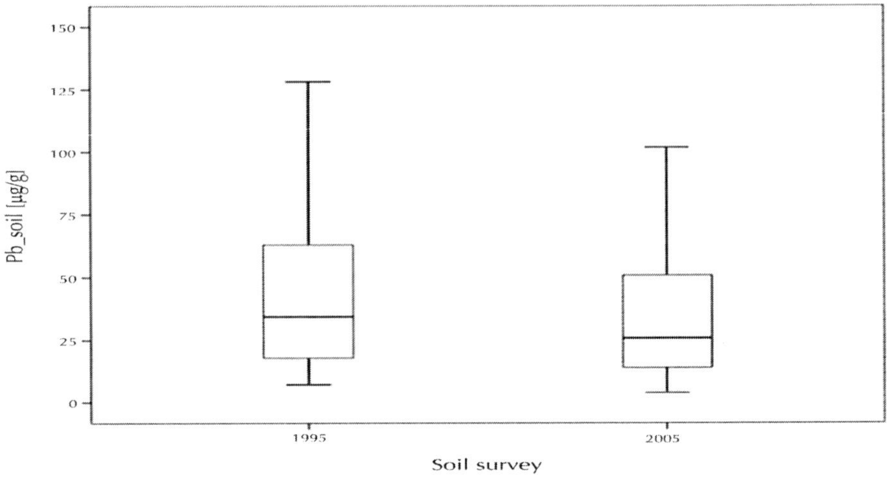

Figure 6: Box plot for Pb in soil Norway 1995 and 2005 (no outliers depicted).

Correlations between HM Concentrations in Moss and Soil Samples and Potential Predictors

As can be seen from the results of Spearman correlation analysis (Table 3), significant statistical associations exist between measured HM concentrations in moss and soil (alpha =0.01). Highest coefficients were evident for Pb (R >0.8), followed by Cd (R>≈0.5). Correlations between Hg concentrations in moss and soil samples ranged between R≈0.3 and R≈0.4.

Table 3: Spearman correlations between measured Cd, Hg and Pb concentrations in moss and soil samples, Norway

	Number	Cd	Pb	Hg
Moss 1990 - soil 1995	211	0.59	0.87	
Moss 1995 - soil 1995	456	0.65	0.84	
Moss 2000 - soil 1995	463	0.57	0.78	
Moss 2000 - soil 2005	219	0.50	0.73	0.29
Moss 2005 - soil 2005	364	0.56	0.76	0.36

Moss 2010 - soil 2005	462	0.56	0.72	0.11

The correlations between the measured HM concentrations in moss samples collected in 1990 with the measured HM concentrations in soil specimen sampled in 1995 included only those soil sampling sites with a spatial distance of 2 km at maximum. The sampling sites for moss and soil were identical in 1995 as well as in 2005. The italicized values denote $p < 0.01$.

Meyer *et al.*

Meyer *et al. Environmental Sciences Europe* 2014 26:27 doi:10.1186/s12302-014-0027-0

Table 4 depicts the results of Spearman correlation analysis of, on the one hand, measured Cd, Hg and Pb concentrations either in moss and in soil samples and, on the other hand, potential influencing factors such as sea spray effect, elevation, precipitation, land use, population density, soil texture (percentage of clay, silt and sand) and total atmospheric deposition of Cd, Hg and Pb modelled by Environmental Monitoring and Evaluation Programme (EMEP). Regarding the element concentrations in moss (Table 4), the modelled deposition values by far showed the highest significant (alpha =0.01) correlations especially for Pb ($R > \approx 0.7$ for all years) and Cd ($R > \approx 0.6$ for all years). Like in the case of natural surface soil (Table 4), for Hg, the weakest correlations among all three elements were identified, ranging from $R \approx 0.3$ in 2010 and $R \approx 0.5$ in 1990. The same tendency can be observed regarding the statistical association of the metal concentrations in moss and the urban land use percentages within 1- and 5-km radiuses (Table 4): Here, the corresponding coefficients are lower ($R < \approx 0.5$) in all cases. The percentages of agriculture land use ($R_{max} \approx 0.2$ for Pb in 1995) and forest areas ($R_{max} \approx 0.3$ for Cd in 1995) showed generally low and mostly non-significant correlations with the HM concentrations in moss. For agricultural, forested and urban land use, the highest associations were identified within a radius of 5 km. Regarding elevation, mostly low negative ($R < \approx -0.3$) and in six cases non-significant coefficients were computed. Similar holds true for the distance between monitoring sites and the sea: Only for Hg, a weak significant signal for all campaigns except for 2010 was observed.

Table 4: Spearman correlations between Cd, Hg, and Pb concentrations in moss and soil and environmental predictors

Descriptor	1990			1995			2000			2005			2010		
	Cd	Hg	Pb	Cd	Hg	Pb	Cd	Hg	Pb	Cd	Hg	Pb	Cd	Hg	Pb
Moss															
Distance to sea (km)	0.11	−0.29	0.01	0.09	−0.22	−0.05	−0.03	−0.29	−0.11	0.06	−0.24	−0.07	0.00	−0.08	−0.15
Elevation (m)	−0.01	−0.30	−0.09	−0.05	−0.14	−0.14	−0.15	−0.31	−0.19	−0.04	−0.22	−0.15	−0.12	−0.13	−0.22
Precipitation (mm)	0.10	0.22	0.25	0.11	0.10	0.22	0.07	0.06	0.23	0.10	0.17	0.19	0.10	0.14	0.21
Ratios agricultural land use 1 km (%)	0.04	0.05	0.07	0.08	0.08	0.12	0.09	0.11	0.11	0.06	0.21	0.12	0.09	0.12	0.13
Ratios agricultural land use 5 km (%)	0.14	0.14	0.22	0.17	0.20	0.23	0.15	0.12	0.17	0.14	0.20	0.18	0.15	0.09	0.18
Ratios forestal land use 1 km (%)	0.11	−0.13	0.13	0.18	−0.12	0.10	0.11	−0.06	0.07	0.15	−0.07	0.07	0.10	0.04	0.06
Ratios forestal land use 5 km (%)	0.27	−0.09	0.28	0.33	−0.08	0.27	0.25	0.01	0.22	0.28	−0.01	0.22	0.25	0.09	0.21
Ratios urban land use 1 km (%)	0.21	0.19	0.24	0.20	0.12	0.25	0.19	0.16	0.22	0.13	0.02	0.19	0.11	0.05	0.14
Ratios urban land use 5 km (%)	0.27	0.25	0.35	0.27	0.21	0.36	0.28	0.14	0.31	0.34	0.06	0.35	0.32	0.12	0.34

Population density (residents/ km2)	0.35	0.39	0.46	0.34	0.42	0.48	0.34	0.31	0.44	0.36	0.31	0.44	0.41	0.17	0.45
Percentage clay in soil (%)	0.12	−0.08	0.05	0.12	−0.04	0.06	0.09	−0.08	0.03	0.19	−0.02	0.10	0.15	0.00	0.10
Percentage silt in soil (%)	−0.15	0.07	−0.07	−0.15	0.03	−0.09	−0.08	0.02	−0.07	−0.12	0.01	−0.03	−0.10	−0.08	−0.05
Percentage sand in soil (%)	0.15	−0.07	0.07	0.15	−0.03	0.09	0.08	−0.02	0.07	0.12	−0.01	0.04	0.10	0.08	0.05
Total deposition Cd - 3-year sum (mg/m2/ year)a	0.62			0.68			0.63			0.61			0.73		
Total deposition Hg - 3-year sum (mg/m2/ year)a		0.46			0.43			0.33			0.33			0.26	
Total deposition Pb - 3-year sum (mg/m2/ year)a			0.81			0.82			0.78			0.69			0.72
Soil															
Distance to sea (km)				−0.02		−0.01				0.00	−0.08	0.03			
Elevation (m)				0.02		−0.06				0.07	−0.13	0.00			
Precipitation (mm)				0.18		0.27				0.18	−0.03	0.19			

Ratios agricultural land use 1 km (%)				−0.04		0.07				0.00	0.11	0.05			
Ratios agricultural land use 5 km (%)				−0.03		0.17				0.01	0.14	0.10			
Ratios forestal land use 1 km (%)				0.03		0.10				0.07	0.02	0.08			
Ratios forestal land use 5 km (%)				0.07		0.21				0.14	0.13	0.20			
Ratios urban land use 1 km (%)				0.12		0.21				0.02	0.08	0.08			
Ratios urban land use 5 km (%)				0.13		0.29				0.20	0.20	0.29			
Population density (inhabitants/km2)				0.23		0.45				0.25	0.28	0.40			
Percentage clay in soil (%)				−0.07		0.00				0.05	0.10	0.07			
Percentage silt in soil (%)				−0.02		−0.05				0.06	−0.03	−0.03			
Percentage sand in soil (%)				0.01		0.01				0.03	0.02	0.02			

Total deposition Cd - 3-year sum (mg/m2/ year)a				0.60						0.51					
Total deposition Hg - 3-year sum (mg/m2/ year)a											0.16				
Total deposition Pb - 3-year sum (mg/m2/ year)a						0.81						0.68			

[a]With exception to 1990 where only modelled data from 1990 was available. Italicized values signify p <0.01.

Meyer *et al.*

Meyer *et al.* *Environmental Sciences Europe* 2014 26:27 doi:10.1186/s12302-014-0027-0

In a similar way as for the HM concentrations in moss, HM concentrations in soil samples were preferentially correlated with atmospheric HM deposition with *R* values for Cd>≈0.5 and for Pb>≈0.7 (alpha =0.01) (Table 4. Population density was significantly correlated with Pb concentrations (*R* coefficients>≈0.4, alpha =0.01) and with Cd concentrations (*R*≈0.20); thus, compared to atmospheric deposition values, it is clearly lower. Land use within a 5-km radius around moss and soil sampling sites was most strongly correlated with concentrations of Cd (*R*≈0.2 in 2005) and Pb (*R*≈0.3).

Geostatistical Analysis and Mapping

Figures 7, 8, 9, 10, 11 and 12 illustrate the results of variogram analyses, cross-validation and mapping via Lognormal Universal Kriging for Cd, Hg and Pb concentrations in moss and soil specimens sampled in 1995 and 2005. Respective results for Cd, Hg and Pb concentrations in moss collected in 1990, 2000 and 2010. Regarding the element concentrations in moss, in nearly all cases, the modelled variograms show clear spatial autocorrelation patterns with nugget-to-sill ratios lying between 11% (for Hg in 1990) to 55% (for Hg in 2005). Regarding the HM concentrations in the soil samples, the nugget-to-sill ratio ranged between 37% (for Cd in 1995) and 72% (for Pb in 2005). For HM concentration in the moss samples, the cross-validation yielded mean error (ME) close to 0, indicating the semi-variogram models and the following surface estimation via Kriging to be unbiased. This is also the case for Cd and Hg concentrations in soil samples, but not for Pb. Moreover, the median percental errors (MPEs) reach from 18% (Hg in 1995) to 36.4% (Cd in 1990), indicating a fair quality of estimation. For HM concentrations in soil specimen, MPEs reaching from 22% (Hg in 2005) to 38.3% (Cd in 2005) were calculated.

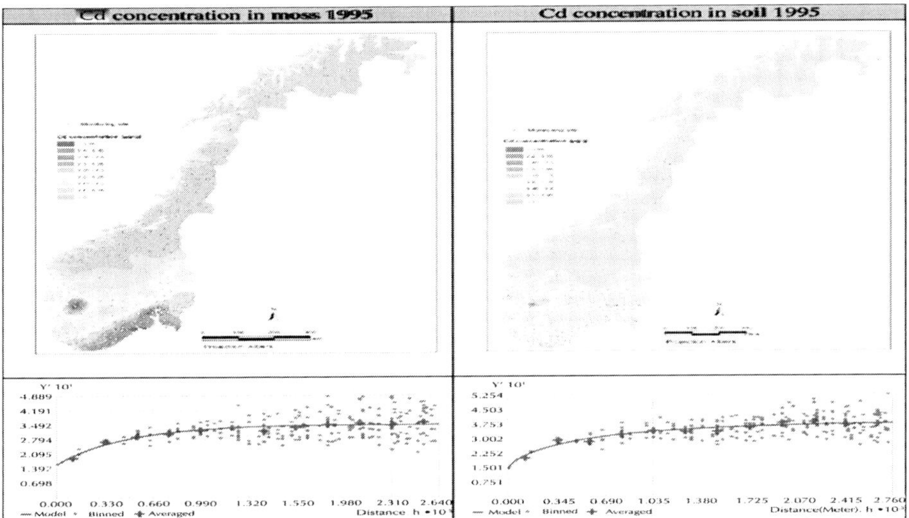

	Cd in moss 1995	Cd in soil 1995
Method	Universal Lognormal Kriging 2nd order	Universal Lognormal Kriging 2nd order
Nugget/Sill Ratio	45 %	37 %
Range	22 km	23 km
Mean Error	0.00341	-0.0137
Median Percental Error	31.4 %	36.0%
Model Type	Exponential	Stable

Figure 7: Geostatistical analysis and mapping of Cd in moss and soil 1995.

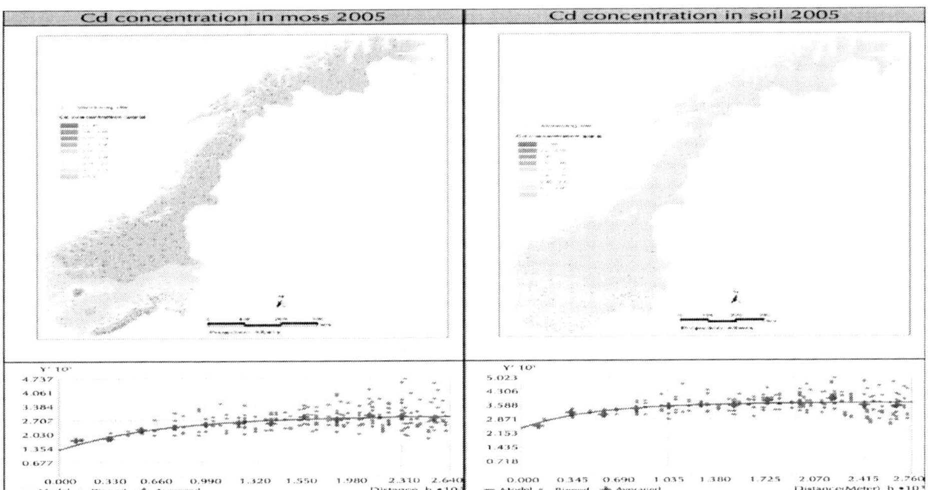

	Cd in moss 2005	Cd in soil 2005
Method	Universal Lognormal Kriging 2nd order	Universal Lognormal Kriging 2nd order
Nugget/Sill Ratio	44 %	65 %
Range	22 km	23 km
Mean Error	-0.00260	-0.00260
Median Percental Error	26.2 %	38.3 %
Model Type	Exponential	Exponential

Figure 8: Geostatistical analysis and mapping of Cd in moss and soil 2005.

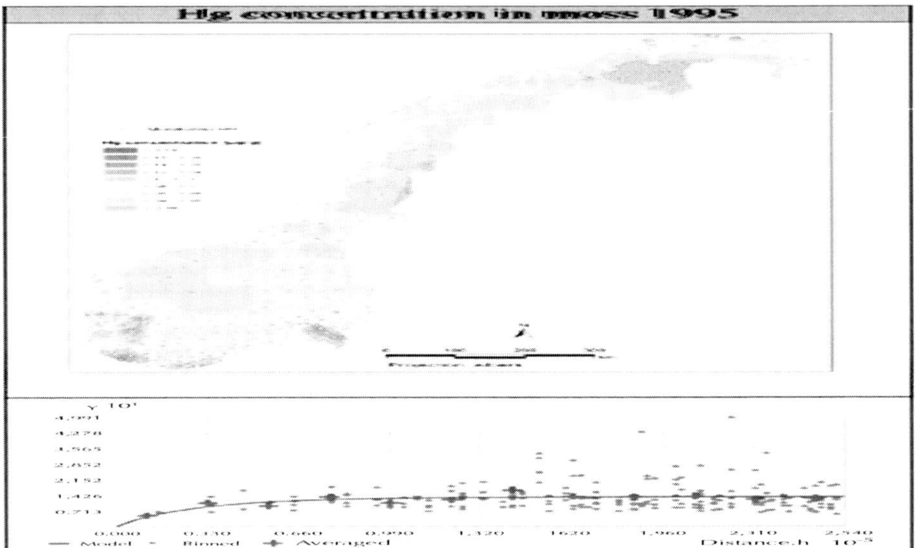

Method: Universal Lognormal kriging 2nd order

Nugget/Sill Ratio	0%
Range	22 km
Mean Error	-0.00063
Median Percental Error	17.9%
Model Type	Stable

Figure 9: Geostatistical analysis and mapping of Hg in moss and soil 1995. Data of Hg concentration in soil 1995 was not available.

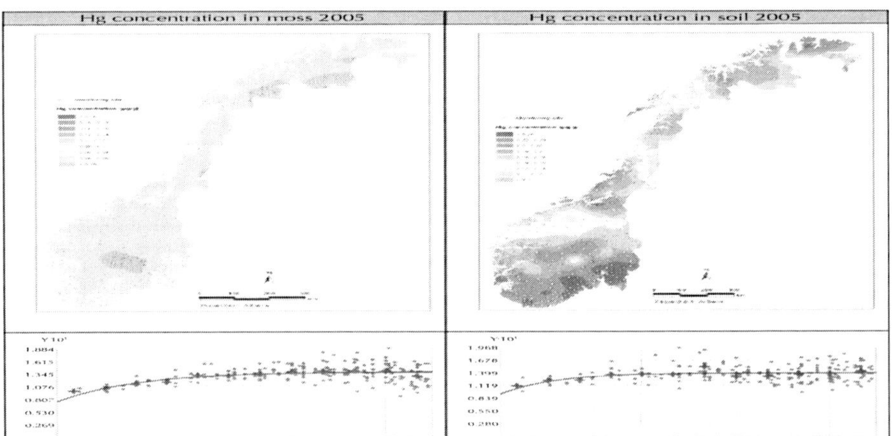

Method: Universal Lognormal Kriging 2nd order

Nugget/Sill Ratio	55%
Range	22km
Mean Error	-0.00003
Median Percental Error	21.3 %
Model Type	Exponential

Method: Universal Lognormal Kriging 2nd order

Nugget/Sill Ratio	67%
Range	16km
Mean Error	0.00016
Median Percental Error	22.0 %
Model Type	Exponential

Figure 10: Geostatistical analysis and mapping of Hg in moss and soil 2005.

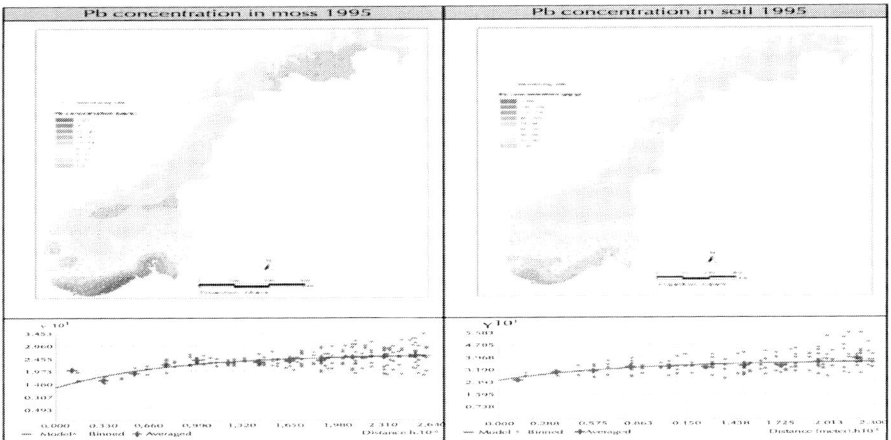

	Pb concentration in moss 1995	Pb concentration in soil 1995
Method	Universal Lognormal Kriging 2nd order	Universal Lognormal Kriging 2nd order
Nugget/Sill Ratio	51%	67%
Range	22km	23km
Mean Error	0.002847	-1.85477
Median Percental Error	27.7 %	35.4 %
Model Type	Exponential	Exponential

Figure 11: Geostatistical analysis and mapping of Pb in moss and soil 1995.

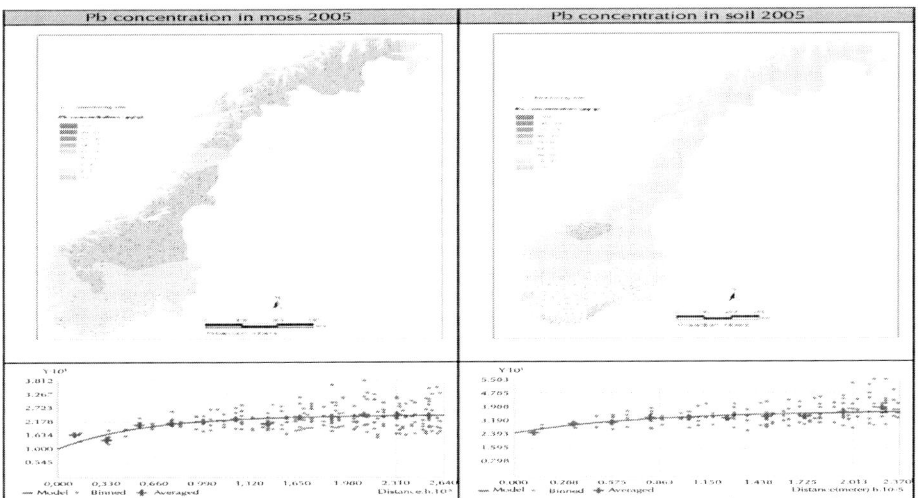

	Pb concentration in moss 2005	Pb concentration in soil 2005
Method	Universal Lognormal Kriging 2nd order	Universal Lognormal Kriging 2nd order
Nugget/Sill Ratio	47%	67%
Range	22km	23km
Mean Error	-0.06968	-1.85477
Median Percental Error	25.3 %	35.4 %
Model Type	Exponential	Exponential

Figure 12: Geostatistical analysis and mapping of Pb in moss and soil 2005.

The Kriging maps (Figures 7, 8, 9, 10, 11 and 12) depict the spatial patterns of the metal concentration in moss for the surveys 1990, 1995, 2000, 2005 and 2010, showing lower metal loads in 2010 than in

1990. For Cd, the highest values are found in the southern part of Norway near the North Sea and Skagerrak coast. In addition, there is a distinct hot spot in the area east of Bergen, related to emissions from the Odda zinc smelter. Pb shows a pattern in Norway similar to that of Cd but without distinct hot spots indicating contributions from local point sources. In 2000, Hg reveals spatial patterns that are very similar to those of Cd and Pb. In 2010, some enrichment is, e.g. depicted in the southern part of Norway around Kristiansand.

The spatial patterns of Cd and Pb concentrations in surface soil are roughly similar to those in moss. With higher values near the North Sea and Skagerrak coast, the maps depicting the spatial patterns of Cd and Pb concentrations in soil samples are spatially less differentiated than those for respective HM concentration in moss samples. Contrary to that is the high spatial differentiation of Hg concentrations in soil sampled in 2005. The maps spatially illustrate that the decrease of HM concentrations is more pronounced in moss than in soil. Soil responds more slowly to decreasing atmospheric HM deposition than moss. This is due to the higher complexity of the regulatory processes acting in soils compared to those in moss. As the spatial extent or the structural und functional complexity of an ecological system increases, so does the temporal extent over which the system's dynamics can be observed [26]-[29].

Table 5 contains the Spearman correlation coefficients between the *measured* Cd, Hg and Pb concentration in moss and the respective Kriging *estimated* concentrations in soil and between the*measured* Cd, Hg and Pb concentration in soil and the respective Kriging *estimated* concentration in mosses (1995 and 2005). Significant statistical associations between HM concentration in moss and soil are evident (alpha = 0.01): The highest coefficients are shown for Pb ($R > \approx 0.8$), followed by Cd ($R > 0.6$ with a maximum of $R \approx 0.78$). Hg shows only moderate correlation patterns. Compared to the results listed in Table 4, the correlation coefficients show the same tendency and are slightly higher.

Table 5: Spearman correlations between measured and geostatistically estimated HM concentrations both in moss and soil

	Number	Cd	Hg	Pb
Moss data and estimated soil data				
Moss 1995 - soil 1995	458	0.78		0.88
Moss 2005 - soil 2005	464	0.74	0.39	0.81
Soil data and estimated moss data				
Soil 1995 - moss 1995	464	0.65		0.86
Moss 2005 - soil 2005	464	0.65		0.86

Values in italics signify $p < 0.01$.

Meyer *et al.*

Meyer *et al. Environmental Sciences Europe* 2014 26:27 doi:10.1186/s12302-014-0027-0

In previous publications from the authors dealing with data from the nationwide surveys [30],[31], the distinctly higher contents of Pb and Cd in the south of the country than elsewhere and the apparent similarity with the geographical distribution in moss samples have led to the conclusion that long-range atmospheric transport from elsewhere in Europe may be the main reason for the observed geographical distributions of these elements in the natural surface soils. The strong statistical asso ciations demonstrated here strongly support this view. In the case of Hg, more uniform distribution over the country may indicate significant contribution from the hemispheric pool of Hg^0, which is assumed to be quite uniformly distributed over the entire northern hemisphere [32].

CONCLUSIONS

The results of this study show that there are significant correlations between HM concentrations in moss and soil collected in Norway during 1990 to 2010 (Table 3). Variogram analysis proved the HM concentrations in moss and soil to be spatially auto-correlated with nugget-to-sill ratios between 11% (for Hg in 1990) to 55% (for Hg in 2005) in moss and between 37% (for Cd in 1995) and 72% (for Pb in 2005) in natural surface soil. In all cases, the MEs are close to 0, indicating that the semi-variogram models and the following Kriging estimation to be unbiased. The MPEs reach from 18% (Hg in 1995) to 36.4% (Cd in 1990) in moss and from 22% (Hg in 2005) to 38.3% (Cd in 2005) in soil, indicating a fair quality of estimation. These results enabled geostatistical surface estimations (Figures 7, 8, 9, 10, 11 and 12). Geostatistical surface estimation allowed to spatially link the HM concentration values with surface data on potential influencing factors. By this, pronounced correlations between Cd, Hg and Pb concentration in moss and soil and influence factors could be established. Following the hypothesis that ecosystems are complex indicators for combinations of causal factors [3], future research is needed to analyse relations between landscape-specific factors and the concentrations of a broader range of metals in moss and soil samples.

METHODS

Moss Sampling and Chemical Analyses

Samples of the feather moss *Hylocomium splendens* (Hedw.) Schimp were collected at around 500 sites distributed over the mainland of Norway and the associated archipelago during the summer of 1977, 1985, 1990, 1995, 2000, 2005 and 2010. Starting in 1995, a fixed network of 464 sites was employed. Samples were dried at room temperature and the last 3 years' growth was selected for determination of trace elements. In 1977 and 1985, the analyses were carried out by instrumental neutron activation analysis for 24 elements and atomic absorption spectrometry for Cu, Zn, Cd and Pb [33]. In 1990 and later years, to which the study at hand relates, inductively coupled plasma mass spectrometry (ICP-MS) was used for all elements, except for Hg which was determined by atomic fluorescence spectrometry. In the 2010 survey, as much as 52 elements were reported [18]. Prior to the ICP-MS analysis, the moss samples were decomposed in a mixture of concentrated nitric acid and hydrogen peroxide and analysed after appropriate dilution. For quality control, three moss reference samples [34] were employed.

Soil Sampling and Chemical Analyses

Surface soil specimens were collected in 1995 and 2005 according to a procedure described by[31]. At most sites, podzols with a well-developed humic surface horizon were sampled. The material collected from the humus layer at 0- to 3-cm depth after removal of litter was analysed by ICP-MS in a similar way as for the moss samples. No data for Hg were obtained in the 1995 soil survey.

Temporal Trend Analyses

Basic descriptive statistical measures (number of sites, minimum, maximum, median) for the Cd, Hg and Pb concentrations in moss were calculated from the moss data collected in 1990, 1995, 2000, 2005 and 2010 as well as from the soil values in 1995 and 2005. In order to depict temporal trends, box plots were derived by use of IBM SPSS Statistics Version 21. To investigate whether significant differences between the data from different monitoring campaigns exist, Wilcoxon signed-rank test were applied.

Correlation Analyses of Measured HM Concentrations

The statistical dependences between the Cd, Hg and Pb concentrations in moss and soil were computed by means of correlation analyses according to [35]. Since soil data were available only for the years 1995 and 2005, HM concentrations in mosses collected from 1990, 1995 and 2000 were compared to the HM concentrations in soil specimen from 1995, whereas the HM concentrations in moss sampled in 2000, 2005 and 2010 were compared to the HM concentrations in soil samples from 2005. The correlations between the measured HM concentrations in moss samples collected in 1990 with the measured HM concentrations in soil specimen sampled in 1995 included only those soil sampling sites which were situated no more than 2 km away from the moss collection sites. The sampling sites for moss and soil were identical in 1995 as well as in 2005. The correlation analyses based on measured HM concentrations in moss and soil samples complement the correlation analyses relying on geostatistically estimated HM concentrations explained in 'Correlation analyses of measured and geostatistically estimated HM concentrations' section below.

Integration of Surface Data on Potential Predictors

Geodata on potential influencing factors were intersected with the moss and soil data in order to explain spatial and temporal variations of Cd, Hg and Pb concentrations in moss and in soil: elevation (GLOBE [36] 1 km × 1 km), precipitation ([37] 20 km × 20 km) and soil texture, i.e. sand, clay and silt content in the upper soil ([38] 1 km × 1 km). The sea spray effect was accounted for in terms of the distance of each moss and soil sampling site to the Norwegian Sea calculated within ArcGIS 10.0. To include land use patterns into the analysis percentages of forests as well as urban and agricultural areas around 1 and 5 km, the moss and soil sampling sites were derived from the Corine Land Cover maps 2000 and 2006 [39] and intersected with the HM concentration values measured in moss and soil specimen. Land use indices derived from the land cover maps 2000 were thereby assigned to the moss data from 1990, 1995 and 2000; those derived from the corresponding map 2006 were assigned to moss data from 2005 and 2010. Furthermore, grid data in a 2.5 arc minute resolution on estimated population densities for the years 1990, 1995, 2000, 2005, 2010 and 2015 were downloaded from the *Gridded Population of the World* (GPW, version 3[39],[40]) and included in the analysis.

Finally, the HM concentrations in moss and soil samples were intersected with maps on modelled total deposition of Cd, Hg and Pb provided by the MSC-E (Meteorological Synthesizing Centre-East) of the EMEP. Within EMEP, the deposition values are modelled for 50 km × 50 km by use of the MSCE-HM atmospheric transport model [41]-[43]. For the metal concentrations in moss, the median for each of the EMEP raster cells was used for the correlation analysis. With exception of 1990, the 3-year sum of the modelled deposition values, preceding the time of the sampling of the moss specimen, were calculated and assigned to each EMEP cell since the analysed moss shoots represent the recent 3 years of growth. For 1990 moss values, only modelled deposition data from that same year was available.

Variogram Analysis and Kriging Estimation

To investigate spatial patterns and temporal trends of the measured Cd, Hg and Pb concentrations in moss (1990, 1995, 2000, 2005 and 2010) and in soil (1995, 2005) and to relate them spatially, geostatistics were applied. To this end, variogram analysis and Kriging procedures were carried through by use of the ArcGIS 10.1 extension *Geostatistical Analyst*. The average nearest neighbour distance was calculated for each year of the abovementioned sampling campaigns and used as a starting point to set the lag distances, amounting for about 22 km in each case. Since all element concentrations showed highly right-skewed data distributions and clear spatial drifts, Lognormal Universal Kriging was applied to calculate surface maps for Cd, Hg and Pb concentration in Norwegian moss and soil samples in a spatial resolution of 5 km × 5 km. Within a four-sector searching window, the ten sites nearest to each raster cell were used to perform the surface estimations in each case. To assess the quality of estimation, the ME and the MPE were calculated from the results of cross-validation.

Correlation Analyses of Measured and Geostatistically Estimated HM Concentrations

Next to mapping spatiotemporal trends, the surface maps calculated by use of the Kriging estimation were used to analyse the statistical dependencies between both the measured Cd, Hg and Pb concentration in moss and the spatially estimated soil data, and the measured Cd, Hg and Pb concentration in soil and the spatially estimated moss data in 1995 and 2005. This procedure complements the correlation analyses based on the measured HM concentrations explained in the 'Correlation analyses of measured HM concentrations' section.

AUTHORS' CONTRIBUTIONS

ES and HU supplied the data. WS headed the investigation, and MM and RP executed the computations. All authors participated in writing the article and read and approved the final manuscript.

REFERENCES

1. Breton J, Massart S, Vandamme P, De Brandt E, Pot B, Foligné B (2013) Ecotoxicology inside the gut: impact of heavy metals on the mouse microbiome. BMC Pharmacol Toxicol 14(62):1-11
2. Harmens H, Norris D, Mills G (2013) Heavy Metals and Nitrogen in Mosses: Spatial Patterns in 2010/2011 and Long-Term Temporal Trends in Europe. Bangor, ICP Vegetation Programme Coordination Centre.
3. Mayack DT (2012) Hepatic mercury, cadmium, and lead in mink and otter from New York State: monitoring environmental contamination. Environ Monit Assess 184:2497-2516
4. Nagajyoti PC, Lee KD, Sreekanth TVM (2010) Heavy metals, occurrence and toxicity for plants: a review. Environ Chem Lett 8:199-216
5. Stankovic S, Kalaba P, Stankovic AR (2014) Biota as toxic metal indicators. Environ Chem Lett 12:63-84
6. (2007) Health Risks of Heavy Metals from Long-Range Transboundary Air Pollution. Joint WHO/Convention Task Force on the Health Aspects of Air Pollution. WHO Regional Office for Europe, Copenhagen.
7. Hettelingh JP, Sliggers J, van het Bolcher M, Denier van der Gon H, Groenenberg BJ, Ilyin I, Reinds GJ, Slootweg J, Travnikov O, Visschedijk A, de Vries W (2006) Heavy Metal Emissions, Depositions, Critical Loads and Exceedances in Europe. PBL, The Hague.
8. Blagnyté R, Paliulis D (2010) Research into heavy metals pollution of atmosphere. Applying moss as bioindicator: a literature review. Environ Res Eng Manag 54(4):26-33
9. Burton MAS (1990) Terrestrial and aquatic bryophytes as monitors of environmental contaminants in urban and industrial habitats. Bot J Lin Soc 104:267-286
10. Rühling A, Tyler G (2001) Changes in atmospheric deposition rates of heavy metals in Sweden. Water Air Soil Pollut Focus 1:311-323
11. Tyler G (1990) Bryophytes and heavy metals: a literature review. Bot J Linn Soc 104:231-253

12. Frahm JP (1998) Moose als Bioindikatoren. Quelle & Meyer GmbH & Co, Wiesbaden.
13. Gecheva G, Yurukova L (2014) Water pollutant monitoring with aquatic bryophytes: a review. Environ Chem Lett 12:49-61
14. Makholm MM, Mladenoff DJ (2005) Efficacy of a biomonitoring (moss bag) technique for determining element deposition trends on a mid-range (375 km) scale. Environ Monit Assess 104:1-18
15. Steinnes E, Berg T (1997) Use of mosses (*Hylocomium splendens* and *Pleurozium schreberi*) as biomonitors of heavy metal deposition: from relative to absolute deposition values. Environ Pollut 98(1):61-71
16. Zechmeister HG, Dullinger S, Hohenwallner D, Riss A, Hanus-Illnar A, Scharf S (2006) Pilot study on road traffic emissions (PAHs, heavy metals) measured by using mosses in a tunnel experiment in Vienna, Austria. Environ Sci Pollut Res 13(6):398-405
17. McGeer J, Henningsen G, Lanno R, Fisher N, Sappington K, Drexler J (2004) Issue Paper on the Bioavailability and Bioaccumulation of Metals. Eastern Research Group, Inc, Lexington.
18. Steinnes E, Berg T, Uggerud H (2011) Three decades of atmospheric metal deposition in Norway as evident from analysis of moss samples. Sci Total Environ 412–413:351-358
19. Rühling A, Tyler G (1973) Heavy metal deposition in Scandinavia. Water Air Soil Pollut 2:445-455
20. Steinnes E, Rambæk JP, Hanssen JE (1992) Large scale multi-element survey of atmospheric deposition using naturally growing moss as biomonitors. Chemosphere 25(5):735-752
21. Harmens BA, Büker P, Norris D, Mills G, Wiliams B, Reynolds B, Ashenden TW, Rühling A, Steinnes E (2004) Heavy metal concentrations in European mosses: 2000/2001 survey. J Atmos Chem 49:425-436
22. Harmens H, Norris DA, Koerber GR, Buse A, Steinnes E, Rühling A (2008) Temporal trends (1990-2000) in the concentration of cadmium, lead and mercury in mosses across Europe. Environ Pollut 151:368-376
23. Harmens H, Norris DA, Koerber GR, Buse A, Steinnes E, Pisspanen J, Alber L, Aleksiayenak Y, Blum O, Coskun M, Dam

M, De Temmerman L, Fernández JA, Frolova M, Frontasyeva M, Gonzalez Miqueo L, Grodzinska K, Jeran Z, Korzekwa S, Krmar M, Kvietkus K, Leblond S, Liiv S, Magnusson SH, Mankovska B, Pesch R, Rühling Å, Santamaria JM, Schröder W, Spiric Z, et al. (2010) Mosses as biomonitors of atmospheric heavy metal deposition: spatial patterns and temporal trends in Europe. Environ Pollut 158:144-156

24. Steinnes E, Berg T, Uggerud H, Pfaffhuber KA (2011) Deposition of heavy metals around Norwegian industries studied by analysis of moss: survey in 2010. In: State Program for Pollution Monitoring, Report 1110/2011. Climate and Pollution Agency, Oslo. pp 1-75 (in Norwegian)
25. Aboal JR, Fernandéz JA, Boquete T, Carballeira A (2010) Is it possible to estimate atmospheric deposition of heavy metals by analysis of terrestrial mosses? Sci Total Environ 408:6291-6297
26. Bellier E, Monestiez P, Durbec J-P, Candau J-N (2007) Identifying spatial relationships at multiple scales: principal coordinates of neighbour matrices (PCNM) and geostatistical approaches. Ecography 30:385-399
27. Chave J (2013) The problem of pattern and scale in ecology: what have we learned in 20 years? Ecol Lett 16:4-16
28. Tóth G, Gardi C, Bódis K, Ivits É, Aksoy E, Jones A, Jeffrey S, Petursdottir T, Montanarella L (2013) Continental-scale assessment of provisioning soil functions in Europe. Ecol Proc 2(32):1-18
29. Wiens JA (1989) Spatial scaling in ecology. Funct Ecol 3:385-397
30. Steinnes E, Allen RO, Petersen HM, Rambæk JP, Varskog P (1997) Evidence of large scale heavy-metal contamination of natural surface soils in Norway from long-range atmospheric transport. Sci Total Environ 205:255-266
31. Nygård T, Steinnes E, Røyset O (2012) Distribution of 32 elements inorganic surface soils: contributions from atmospheric transport of pollutants and natural sources. Water Air Soil Pollut 223:699-713
32. Steinnes E, Berg T, Sjøbakk TE (2003) Temporal and spatial trends in Hg deposition monitored by moss analysis. Sci Total Environ 304:215-219

33. Steinnes E (1980) Atmospheric deposition of heavy metals in Norway studied by analysis of moss samples using neutron activation analysis and atomic absorption spectrometry. J Radioanal Chem 58:387-391
34. Steinnes E, Rühling Å, Lippo H, Mäkinen A (1997) Reference materials for large-scale metal deposition surveys. Accred Qual Assur 2:243-249
35. Spearman C (1904) The proof and measurement of association between two things. Am J Psychol 15:72-10.
36. Hastings DA, Dunbar PK (1999) Global Land One-Kilometer Base Elevation (GLOBE) Digital Elevation Model, Documentation, Volume 1.0. Key to Geophysical Records Documentation (KGRD) 34. National Geophysical Data Center, National Oceanic and Atmospheric Administration, Boulder.
37. New M, Lister D, Hulme M, Makin I (2002) A high-resolution data set of surface climate over global land areas. Clim Res 21:1-25
38. (2009) Harmonized World Soil Database (version 1.1). FAO/IIASA, Rome/Laxenburg.
39. Büttner G, Kosztra B, Maucha G, Pataki R (2012) Implementation and Achievements of CLC2006. Bellaterra (Barcelona). European Topic Centre and European Environment Agency, Copenhagen.
40. (2005) Gridded Population of the World, Version 3 (GPWv3): Population Count Grid. NASA Socioeconomic Data and Applications Center (SEDAC), Palisades.
41. Aas W, Breivik K (2006) Heavy Metals and POP Measurements, 2006. EMEP/CCC-Report 4/2008. Norwegian Institute for Air Research, Oslo.
42. Ilyin I, Rozovskaya O, Travnikov O, Aas W, Hettelingh JP, Reinds GJ (2008) Heavy Metals: Transboundary Pollution of the Environment. EMEP Status Report 2/2008. Meteorological Synthesizing Centre, East, Moscow.
43. Gusev A, Ilyin I, Rozovskaya O, Shatalov V, Sokovych V, Travnikov O (2010) Modelling of Heavy Metals and Persistant Organic Pollutants: New Developments. EMEP/MSC-E Technical Report 1/2009. Meteorological Synthesizing Centre - East, Moscow.

Chapter 5

Retention and Mitigation of Metals in Sediment, Soil, Water, and Plant of a Newly Constructed Root-Channel Wetland (China) from Slightly Polluted Source Water

Baoling Wang[1,2], Yu Wang[1,2], and Weidong Wang[1]

[1]State Key Laboratory of Environmental Aquatic Chemistry, Research Center for Eco-Environmental Sciences, The Chinese Academy of Sciences, Beijing 100085, P. R. China

[2]University of Chinese Academy of Sciences, Beijing 100049, P. R. China

ABSTRACT

Constructed root-channel wetland (CRCW) is a term for pre-pond/wetland/post-pond complexes, where the wetland includes plant-bed/ditch landscape and root-channel structure. Source water out of pre-ponds flows through alternate small ditches and plant beds with root-channels via a big ditch under hydraulic regulation. Then source water flows into post-ponds to finish final polishing. This article aims to explore the potential of components of a pilot CRCW in China on mitigating metals in micro-polluted source water during its initial operation stage. We investigated six heavy metals (Cd, Cr, Cu, Ni, Zn, and Pb) in surface sediment, plant-bed subsurface soil, water, and aquatic plants during 2012–2013. Monitoring results showed that pond/ditch sediments and plant-bed soil retained a significant amount of Cr, Ni, and Zn with 93.1%, 72.4%, and 57.5% samples showing contamination factor above limit 1 respectively. Remarkably the high values of metal enrichment factor (*EF*) occurred in root-channel zones. Water monitoring results indicated that Ni, Zn, and Pb were removed by 78.5% (66.7%), 57.6% (59.6%), and 26.0% (7.5%) in east (west) wetland respectively. Mass balance estimation revealed that heavy metal mass in the pond/ditch sediments accounted for 63.30% and that in plant-bed soil 36.67%, while plant uptake occupied only 0.03%. The heavy metal accretion flux in sediments was 0.41 - 211.08 $\mu g \cdot cm^{-2} \cdot a^{-1}$, less than that in plant-bed soil (0.73 - 543.94 $\mu g \cdot cm^{-2} \cdot a^{-1}$). The 1.83 ha wetland has retained about 86.18 kg total heavy metals within 494 days after operation. This pilot case study proves that constructed root-channel wetland can reduce the potential ecological risk of purified raw water and provide a new and effective method for the removal of heavy metals from drinking water sources.

INTRODUCTION

Metals, especially heavy metals, have received considerable attention owing to their contamination on aquatic environments. In particular, they are common pollutants in water sources and drinking water, prompting concern regarding their hazardous effects on the environment and the food chain (Graney and Eeriksen 2004; Stead-Dexter and Ward 2004). Moreover, they are known to become concentrated, and diffuse spatially and have been shown to induce adverse effects at physical and biochemical levels, including toxication, carcinogenesis, teratogenesis, and mutagenesis (Weis and Weis 2004; Wu et al. 2013). Many novel technologies and approaches to the treatment of metals have been studied intensively (Rai 2012; Rana et al. 2013). Application of constructed wetland in water treatment has gained popularity throughout the world for its low costs, low energy consumption and minimal operation and maintenance. Constructed wetland can effectively remove heavy metals in water sources and waste water by acting as a matrix, facilitating interaction between microbes and plant/animal communities and performing functions such as filtration, adsorption, precipitation, ionic exchange, microbiological degradation, and biological uptake (Kadlec 2003; Vymazal 2011; Si et al. 2011; Malaviya and Singh 2012). Recent reports have focused primarily on the sources and distributions of heavy metals (Zhang et al. 2002; Protano et al. 2013; Mohammed and Abdu 2013) and on assessing the risks posed by such metals (Xiao et al.2012; Yang et al. 2008; Vymazal and Krása 2003; Obarska-Pempkowiak and Klimkowska 1999; Goulet et al. 2001).

The root-channel wetland here studied is also called constructed root-channel wetland (CRCW), a type of constructed wetland proposed by our research team and currently under patent protection. It is an innovative wetland configuration that includes the macroscopic plant-bed/ditch landscapes and mesoscopic root-channel structures originated from Baiyangdian Lake in North China (Wang et al. 2012a; Wang et al. 2012b). Fundamentally, the CRCW technology utilizes various crop stalks (fresh or dry, thick or thin) as substrates that are buried in the wetland subsurface soil layers during the initial operation stages to form artificial root-channels. Simultaneously, aquatic plants with developed rhizomes and adventitious roots are transplanted gradually to form natural root-channels, which allow the efficient transmission of gases, liquids, and other materials into the deep soil matrix. A successful CRCW case study has been undertaken in the Shijiuyang wetland, located in Yangtze River delta in China (Figure 1). The wetland becomes an effective buffer and processor to the complex pollution of drinking water source which involves the removal of nutrients (Wang et al. 2012b), heavy metals (Wang et al. 2014), polycyclic aromatic hydrocarbons (PAHs) (Zheng et al. 2012b), and algae (Zheng et al. 2012a), as well as the promotion of bacteria community structure (Wang et al. 2011). Investigation of the long-term effects of accumulation of metals, especially heavy metals, has been conducted here (Wang et al. 2014). The results demonstrate that the combination of multilevel ponds and plant-bed/ditch systems can intercept and retain metals from source water. The achievements of this project in terms of ecological drinking water source practices were recognized by the presentation of several awards, including the Best Practices in Human Settlements Improvement in China in 2011 and the Dubai International Award for Best Practices to Improve the Living Environment in 2012.

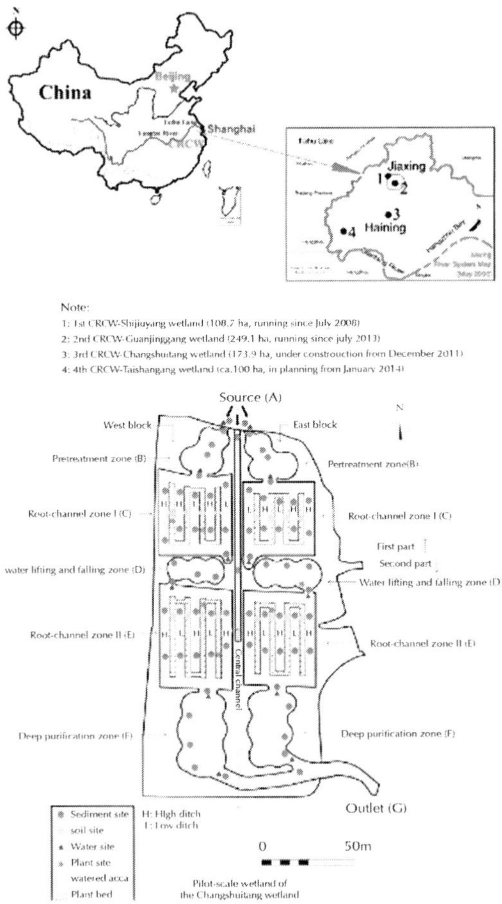

Figure 1: Location map of constructed root-channel wetlands (CRCWs) and sampling sites in the pilot wetland. CRCWs are located in the center of Yangtze River delta on China's southeast coast (top left), and CRCWs 1, 2 in Jiax-

ing City, CRCWs 3, 4 in Haining City (top right). Sediments from ponds and ditches, soils from plant beds, water, and plant above ground portion were sampled in a pilot-scale wetland of CRCW 3 (down).

More recently, the Changshuitang wetland in Haining became the site of the third CRCW (Figure 1). To further optimize the water treatment efficiency of this full-scale wetland (ca. 1.73 km^2), a pilot wetland at a scale of approximately 1:100 (ca. 1.83 ha) was constructed in advance nearby and continued operating for over a year. The goals of the present study for this pilot wetland are: (1) To study the distribution of heavy metals and their accumulation effects on the wetland segments during initial period of pilot wetland, further to provide the guidance for the construction of full-scale wetland later; (2) To gain an enhanced understanding of the mechanism(s) on how the potential ecological health risks through purified raw water can be reduced even at low metal concentration.

RESULTS

Metal Contents and Risk Assessment in Sediments/Soil

Metal Contents

The results of analysis of water demonstrate that the pilot wetland exhibited considerable spatial heterogeneity between water and soil (surface sediment and plant soils) samples after operating for over 1 year, particularly with respect to metals (Table 1). The spatial variations of water and soil quality indices suggests the presence of heterogeneity of water quality purification processes along the hydraulic flow pathways in the wetlands and variability in the enrichment effects for metals in the five functioning zones and structural units of the root-channel zone. For example, in the 84 sediment/soil samples, heavy metal contents were typically lower than those of major metals. The mean coefficient of variation (CV) for all metals in the pilot wetland was found to be 44.90%, with CVs decreasing in the following order:

Ca >> Mg ≈ Na ≈ Zn > Pb > K ≈ Cd > Al > Cr ≈ Ni ≈ Cu >> Fe. The highest metal contents in the sediments of the wetlands exceeded the corresponding levels in source river sediments, suggesting that accumulation and enrichment did indeed occur within the wetlands, at least in some locations. The average contents of heavy metals in the composite source river sediment (Site A) are as follows (mg/kg): Cd: 0.09, Cr: 151.07, Cu: 32.34, Ni: 55.88, Zn: 167.52, Pb: 14.68. The maximum contents of all six heavy metals in the wetland soil/sediments (Sites B–G) exceeded the corresponding source levels at Site A without exception (average ratios: 2.37), with ratios (i.e., wetland/source) as follows: Cd: 3.08, Cr: 1.67, Cu: 1.22, Ni: 1.32, Zn: 2.55, Pb: 4.37. Despite sharing the same source, the metals exhibited considerable spatial variation throughout the wetland: we found that the contents were 0.17–4.62 (mean: 1.05, CV: 49.59%, median: 0.93) times the corresponding background levels in the surface soils of the Hangzhou–Jiaxing–Huzhou plain for heavy metals. Compared with the source river sediments, Cd, and Pb were enriched in the pilot wetland on the whole, although their contents in source river sediments were as low as the background levels. Moreover, contents of Cr, Ni, and Zn in source river sediments were already higher than the Hangzhou–Jiaxing–Huzhou plain background levels adopted here. Heavy metal contents in the sediments of the central channel (i.e., newly constructed channel and deposited sediments) were relatively high (approximately 0.93–2.50 times background levels, mean: 1.28), indicating some degree of enrichment. Sediment particulates settled with distance along the flow pathways in the central channel (typically over dozens of meters). In contrast, contents of heavy metals such as Cu, Ni, Zn, and Pb increased by 11.12%, 17.49%, 16.14%, and 34.03%, respectively, demonstrating that heavy metals were retained primarily in the latter part of the central channel. These results suggest that superficial sedimentation could act as a primary sink for heavy metals in the stream

Table 1: Contents of metals in source river sediments, wetland sediments/soils as compared with regional background levels

Heavy metal	Range mg/kg	Mean mg/kg	CV	Source mg/kg	Background mg/kg	Major metal	Range	Mean	CV	Source	Background
Cd	0.05–0.28	0.12	38.80%	0.09	0.15	K	1.16–8.50%	3.74%	38.83%	1.79%	2.03%
Cr	70.32–251.95	121.57	25.70%	151.07	77.55	Na	0.42–24.28%	8.10%	61.80%	7.37%	1.08%
Cu	12.33–39.33	22.07	21.33%	32.34	30.85	Ca	0.21–12.58%	3.03%	110.06%	4.34%	0.69%
Ni	21.74–73.70	38.93	23.52%	55.88	32.40	Mg	0.22–2.02%	0.70%	63.68%	0.95%	0.92%
Zn	1.96–427.93	112.95	61.64%	167.52	92.67	Al	0.97–16.84%	9.69%	33.25%	6.41%	7.57%
Pb	5.22–64.20	25.37	48.67%	14.68	30.35	Fe	2.45–4.93%	3.76%	11.53%	3.54%	3.23%

Note: $n = 84$; CV: coefficient of variation; Background levels are contents in the surface soils of the Hangzhou–Jiaxing–Huzhou plain (Wang et al. 2007).

Wang *et al.*

Risk Assessment of Heavy Metals

According to Environment Canada standards (MacDonald et al. 2000), two distinct thresholds can be used to assess pollution conditions in wetland sediments: the threshold effect level (TEL) and the probable effect level (PEL). Heavy metal contents lower than TEL are indicative of rare or minor pollution with little to no biological toxicity; thus, such pollution rarely induces negative ecological effects. In contrast, heavy metal contents that lie between TEL and PEL are indicative of moderate pollution that occasionally poses negative ecological risks. When heavy metal content exceeds PEL, serious pollution may occur and may be associated with considerable biotoxicity; in such instances, the negative ecological effects can be pronounced. All sediment samples collected in the pilot wetland exhibited Cd values that were below TEL, suggesting negligible Cd pollution in the wetland. However, the wetland was found to be slightly polluted by Cu, Pb, and Zn, with 98.81%, 88.10%, and 65.48% of sites between TEL and PEL. Cr and Ni contents exceeded PEL in 84.52% and 53.57% of sites. Thus, it can be assumed that heavy metals exhibit a generally low ecological risk in the pilot wetland, although Cr and Ni should be monitored closely. Further details are showed in Table 2.

Table 2: Risk assessment of heavy metals in source river sediment, wetland sediments/soils, and regional background soils (mg/kg, dried weight)

Heavy metals	TEL	PEL	Hangzhou–Jiaxing–Huzhou plain soil layer A	Source river	Within wetland, n=84			Percentage		
					Max	Min	Mean	<TEL	TEL ~ PEL	>=PEL
Cd	0.596	3.53	0.152	0.09	0.28	0.05	0.12	100.0%	0.0%	0.0%
Cr	37.3	90	77.6	151.07	251.95	70.32	121.57	0.0%	15.5%	84.5%
Cu	35.7	197	30.8	32.34	39.33	12.33	22.07	98.8%	1.2%	0.0%
Ni	18	36	32.4	55.88	73.70	21.74	38.93	0.0%	46.4%	53.6%
Zn	123	315	92.7	167.52	427.93	21.96	112.95	65.5%	32.1%	2.4%
Pb	35	91.3	30.4	14.68	64.20	5.22	25.37	88.1%	11.9%	0.0%

Note: TEL: threshold effect level; PEL: probable effect level. The plain text, italics, and boldface regarding the contents represent rare, *occasional*, and frequentnegative ecological risks, respectively. Hangzhou–Jiaxing–Huzhou plain soil layer A data are from (Wang et al. 2007).

Wang *et al.*

Distribution of Metals in Wetland Sediments/ Soils

Spatial Variations of Metal Contents

Comparisons between the east and west blocks, the first and second parts, the five functioning zones along the hydraulic flow pathways, and the three finer scale structural units within both root-channel zones represent the basic characteristics of spatial variations within the wetland and are described here and in the following sections (Figure 2). The spatial distribution of metals in the east and west blocks was found to exhibit a complex pattern, which can be summarized as follows. (1) Cd exhibited obviously high contents in the east wetland, with average contents of 0.14 mg/kg (east) and 0.092 mg/kg (west). (2) Cr, Cu, Ni, and Zn contents were higher in the west wetland than the east, with average Cr contents of 103.51 mg/kg (east) and 141.00 mg/kg (west), average Cu content of 19.92 mg/kg (east) and 24.70 mg/kg (west), average Ni contents of 33.12 mg/kg (east) and 45.53 mg/kg (west), and average Zn contents of 75.67 mg/kg (east) and 152.77 mg/kg (west). (3) Pb contents varied little from east to west in the first part, but were higher in the east in the second part. These results demonstrate that, despite almost identical design and operation conditions in the east and west wetlands, nonparallel variations of metals were detected in wetland sediment, with no synergistic variation in the two successive parts of the pilot wetland.

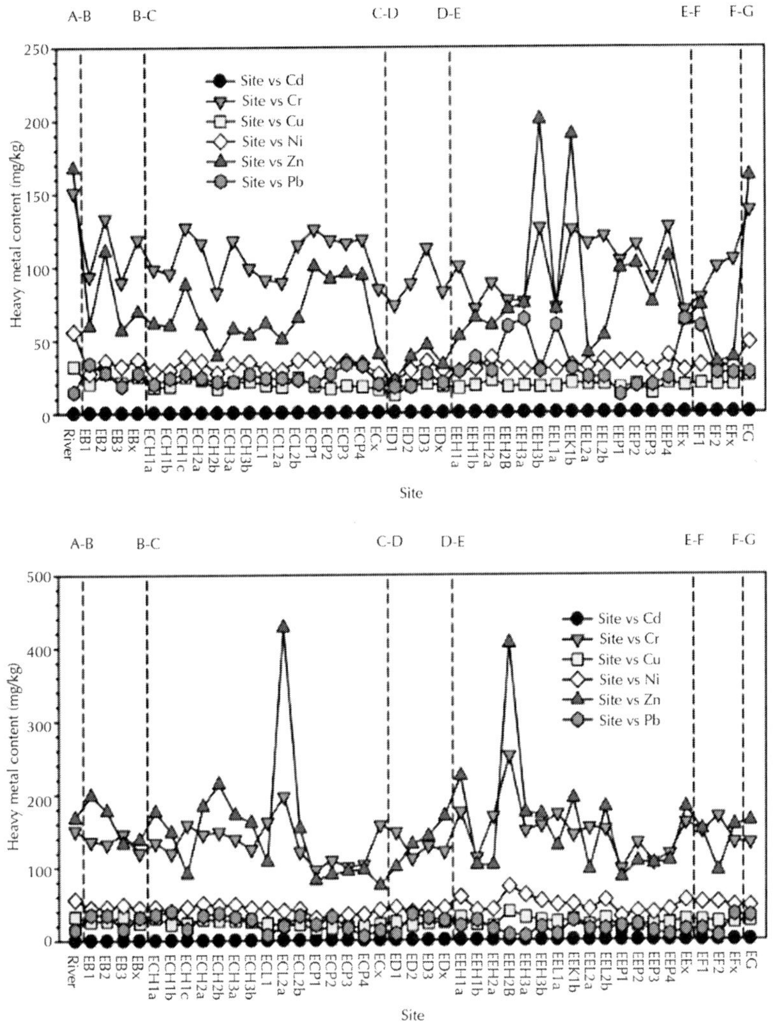

Figure 2: Heavy metal contents in source river and locations of the east (top) and west (down) pilot wetland. Refer to sampling location map in Figure 1. X-axis: River: source river (corresponding to "A" on top X-axis); the first letter "E": east wetland, "W": west wetland; the second letter (corresponding to letters on top X-axis) "B": pretreatment zone, "C": root-channel zone I, "D": water lifting and falling zone, "E": root-channel zone II, "F": deep purification zone, "G": wetland outlet; the third letter "H": high ditch, "L": low ditch, "P": plant bed, "x": the exit of functioning zone; the numbers after "B", "D", "F": locations along hydraulic pathways; the numbers after "H", "L": ditch sequence; the numbers after "P": plant bed sequence; the last letter "a", "b", "c": locations along hydraulic pathways in ditches. Sites on plant beds are for collecting soil and the rest are for sediments.

Spatial Variations of Metal Contamination Factors

Contamination factor and enrichment factors (*EF*s) can be used to indicate metal pollution levels and are more suitable for characterizing retention processes than metal contents. The mean *EF*s of the six heavy metals studied were all considerably lower than 1.50, demonstrating that no obvious heavy metal accumulation (with respect to the low concentrations in the source river water) occurred on the whole during the initial operation period. However, the maximum *EF*s of Cd, Cr, Ni, Zn, and Pb exceeded 1.50, indicating some enrichment of these heavy metals in some parts of the wetland. Contamination factor (*C*) describes the contamination of a given toxic substance. The*C* value represented spatial variations in the pilot wetland. From data in Table 3, we can see contamination factors of most heavy metals in west block are higher than 1, which means the contents of heavy metals in pilot wetland were higher than the corresponding background level. Contamination factors of heavy metal Cd, Cu, Pb in two blocks were lower than 1 and most sites in the pilot had lower metal contents. Cr had 85.71% (average 1.32) and 100.00% (average 1.82) sample numbers that contamination factors were higher than 1 in east and west block, demonstrating the highest accumulation effects. Ni and Zn had similar variations whose contamination factors were higher in west block and more than half sites had contamination factors higher than 1. The maximum contamination factors for heavy metals were all above 1 except Cd in west and Cu in east. The studies proved that the primary operational pilot wetland had relatively accumulated the heavy metals, especially for Cr, Ni and Zn.

Table 3: Statistics and proportion of heavy metal contamination factors (*C*) in wetland sediments/soils

	Proportion (C<1)		Proportion (C>1)		Mean		Max	
	East	West	East	West	East	West	East	West
Cd	66.67%	100.00%	33.33%	0.00%	0.93	0.61	1.83	0.89
Cr	14.29%	0.00%	85.71%	100.00%	1.32	1.82	1.78	3.25
Cu	100.00%	88.10%	0.00%	11.90%	0.64	0.79	0.91	1.27
Ni	52.38%	4.76%	47.62%	95.24%	1.01	1.40	1.48	2.27
Zn	76.19%	11.90%	23.81%	88.10%	0.79	1.64	2.16	4.62

Pb	76.19%	69.05%	23.81%	30.95%	0.96	0.71	2.12	1.31

Note: Proportion means the number percentage of sample sites whose *C* met the demand in brackets accounting for the number of all sample sites.

Wang *et al.*

The heavy metal contamination factors in the pilot wetland. Generally, in east block, contamination factors for Cr, Cu, and Ni were high in pretreatment zone and factors for Cd and Pb were high in deep purification zone. In west block, contamination factors for Cd, Cu, Zn, and Pb were high in pretreatment zone, while factors for Cr, Ni were high in deep purification zone. The four root-channel parts had relatively high contamination factors, but lower than those at the outlets of wetland. Spatial variations of contamination factors in structural units (e.g., high/low water level ditch, plant bed) were investigated to analyze metal accumulation effects in root-channel zones (Table 4). The results reveal differences in contamination factors between the two blocks. In the east block, the contamination factors of Cd, Zn, and Pb in the ditches of the second part were higher than those in the first part, whereas the contamination factors of Cr, Cu, and Ni exhibited no distinct variance. The contamination factors of Cr, Cu, Ni, and Zn exhibited the opposite trend and those of Cd and Pb exhibited no clear variance. The contamination factors of Pb were lower in the plant bed of the second part; the contamination factors of Cd, Cr, Cu, Ni, and Zn exhibited the opposite trend. Moreover, contamination factors in low ditches were often lower than those in high ones.The biplot of metal contamination factors (Figure 3) illustrates the multivariate correlations between variables (metal contamination factors) and observations (structural units). The explanatory percentage of first and second principal components accounted for 68.4% and 25.7%, respectively, explaining a total of 94.1% of the variance. This can be considered a very good fit. The results in Figure 4 demonstrate that no metals exhibited obvious enrichment in the plant beds of the first and second parts of either the east or west wetlands (i.e., all plant beds), or in the high/low ditches in the first part of the east wetland. The heavy metals Pb and Cd exhibited obvious accumulation in the high and low ditches in the second part of the east wetland; such accumulation was also detected for the heavy metals Cr, Cu, Ni, and Zn in the high and low ditches in the second part of the west wetland. In general, the plant-bed/ditch system in the Changshuitang pilot wetland accumulated heavy metals (Cd, Cr, Cu, Ni, Zn, and Pb) primarily in the second part of the wetland, with no apparent effects in the soil of the plant beds.

Table 4: Contamination factors (C) of heavy metals in the structural units of root channel zones

Block	Part	Unit	Cd	Cr	Cu	Ni	Zn	Pb
East	First	High Ditch	0.94±0.19	1.36±0.2	0.67±0.1	1.01±0.13	0.65±0.16	0.78±0.09
		Plant Bed	0.56±0.02	1.54±0.05	0.58±0.03	1.08±0.04	1.03±0.04	0.93±0.19
		Low Ditch	0.76±0.07	1.23±0.17	0.63±0.11	0.94±0.13	0.59±0.12	0.74±0.07
	Second	High Ditch	1.18±0.32	1.16±0.27	0.61±0.06	0.96±0.1	0.94±0.6	1.36±0.54
		Plant Bed	0.65±0.11	1.42±0.19	0.58±0.1	1.07±0.13	1.03±0.15	0.61±0.15
		Low Ditch	1.24±0.25	1.3±0.35	0.63±0.03	0.99±0.08	0.9±0.66	1.34±0.65
West	First	High Ditch	0.71±0.15	1.79±0.17	0.86±0.1	1.42±0.11	1.77±0.41	1.04±0.26

		Plant Bed	0.53±0.06	1.32±0.07	0.55±0.04	1.02±0.07	0.98±0.07	0.61±0.36
		Low Ditch	0.53±0.15	2.06±0.4	0.71±0.02	1.29±0.04	2.06±1.74	0.6±0.37
	Second	High Ditch	0.6±0.14	2.19±0.59	0.92±0.23	1.71±0.39	2.13±1.21	0.51±0.26
		Plant Bed	0.67±0.07	1.47±0.2	0.66±0.09	1.16±0.12	1.1±0.11	0.49±0.21
		Low Ditch	0.58±0.15	2.02±0.14	0.86±0.11	1.54±0.16	1.69±0.45	0.52±0.23

Note: mean ± standard deviation, boldface means data above 1.0.

Wang *et al.*

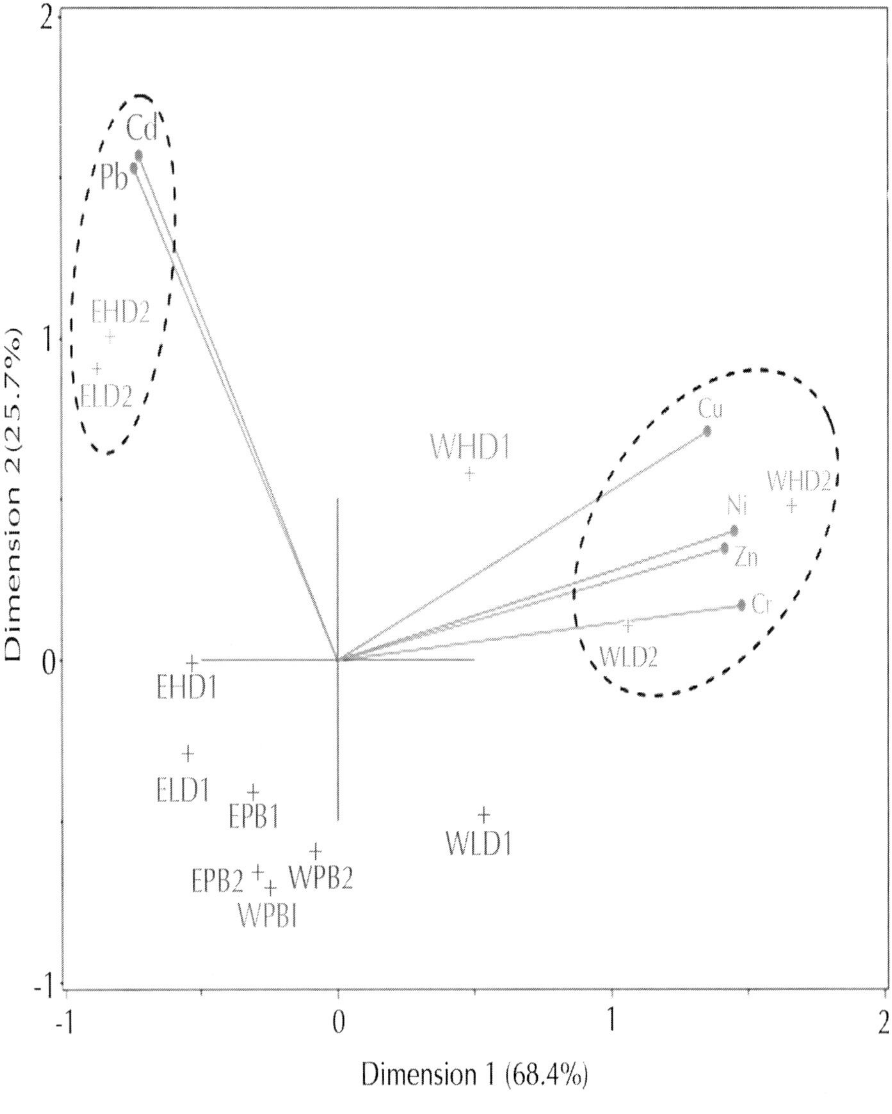

Figure 3: Biplot of heavy metal contamination factors (C) in the plant-bed/ditch systems. EHD, ELD, EPB: high ditch, low ditch, plant bed in east wetland; WHD, WLD, WPB: high ditch, low ditch, plant bed in west wetland.

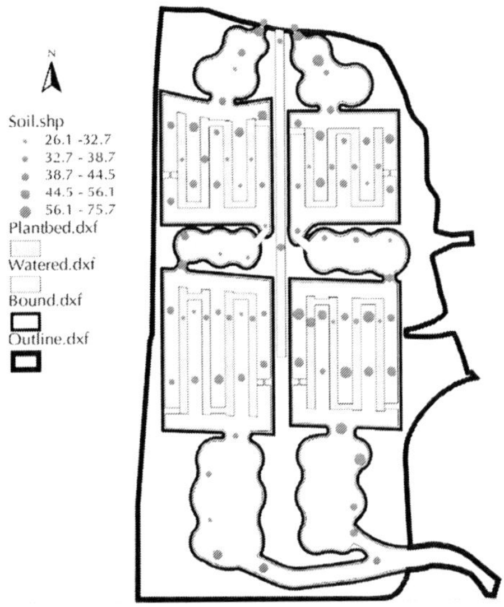

Figure 4: Potential comprehensive ecological risk index (*RI*) of heavy metals in source river sediment, pond/ditch sediments and plant-bed soils of the pilot wetland. Outline, bound line, watered line, plant bed line represent the outer boundary, inner boundary, watered area (light blue), and plant bed area (light green). The solid red circle illustrates the relative magnitudes of *RI*. The size range of circles is classified according to Natural Breaks method in ArcView GIS 3.2a.

Variations of Risk Index (RI) of Heavy Metals

By considering absolute contents and toxic response factors in tandem, the index *E* can indicate the potential ecological risk of an individual heavy metal, whereas the comprehensive potential ecological risk index (*RI*) represents the sum of the potential risk of all individual heavy metals. For individual metal elements in sediment/soil samples, most *E* values were considerably lower than 40, indicating low potential ecological risk (Håkanson 1980). Discernable spatial variations were also found in *RI* of heavy metals (Figure 4): *RI* for the six investigated heavy metals for the entire pilot wetland was calculated to be 26.13–75.74 (mean: 41.18) and was relatively high in the east (mean ± standard deviation, the same hereinafter) (44.39 ± 11.48) but low in the west (37.97 ± 6.69). In the east wetland, *RI* in the second part (47.10 ± 13.16) was higher than that in the first part (41.11 ± 8.25); conversely, in the west wetland, *RI* in the second part (37.81 ± 6.27) was approximately equal to that in the first part (38.17 ± 7.34).

Moreover, it is clear that *RI* exhibits similar patterns in both blocks, with significant decreases in the previous two functioning zones (i.e., the pretreatment and root-channel zones) owing to the effects of settlement in the pretreatment zone and to absorption and accumulation effects in root-channel zones. *RI* decreased gradually along the hydraulic flow pathways, with decreases of 29.93% and 35.70% from inlet to outlet in the first part in the east and west wetlands, respectively. After water pumping, *RI* increased sharply in both high and low ditches, exhibiting increases of 56.29% and 68.40% in the second part of the east and west wetlands, respectively. This demonstrates that both the pre-pond (at the front end of the wetland) and the post-pond (at the back end of the wetland) functioned effectively to retain metals under the push exerted by the water head and hydraulic flow. *RI* in the plant-bed/ditch systems of the root-channel zones exhibited spatial variation, decreasing in the following order: high ditch (45.35 ± 10.19), low ditch (41.09 ± 8.68), plant bed (33.67 ± 3.65). Small ditches within the plant-bed/ditch system in the root-channel zone appeared to cause enrichment by removing metals from source water. To summarize, we found the average *RI* to be higher than the mean of the whole wetland in all zones except the water lifting and falling zone, which had the lowest *RI*. Thus, heavy metals were largely retained along hydraulic flow pathways, helping to mitigate the ecological risk posed by the source water.

Metal Variations in Wetland Water

The concentrations of metals and total suspended solids (TSS) in wetland water samples were measured after sampling. According to Table 5, good removal rates were found for TSS in the pilot wetland, with removal rates of 47.80% and 37.58% in the east and west wetlands, respectively. This was likely the foundation for the removal of metals through deposition and adsorption. Mn in water samples was also measured and its concentrations were found to exceed the standard limited mean (0.10 mg/L) in both the inlet (source water) and the first part of the wetland (max: 0.17 mg/L); however, removal rates of Mn were high, with rates of 70.59% and 42.86% in the east and west wetlands, respectively. Cd, which has the strongest toxicity of the heavy metals examined, was not detected, because the water of the Jujinqiaogang River, which is adjacent to the Changshuitang pilot wetland, was micro-polluted. However, all other heavy metals investigated were detected. In the pilot wetland especially in the east block, concentrations of heavy metals declined sharply from sites B (E2,W2) to C (E3,W3) and from sites D (E4,W4) to E (E5,W5), implying that both root-channel zones in this block achieved good removal of heavy metals. Another obvious decline from sites A (E1,W1) to B (E2,W2) demonstrates that the pretreatment zone also exerted pronounced positive effects through mitigating the metal risk. This decrease corresponds well to changes in metal contents in sediments. Variations in TSS were also found to exhibit a good relationship with those of heavy metals.

Table 5: Heavy metal contents in water column of pilot wetland along water flow pathways

Site	TSS mg/L	Mn mg/L	Cd µg/L	Cr µg/L	Cu µg/L	Ni µg/L	Zn µg/L	Pb µg/L
E1	93.3	0.17	<1	2.27	4.31	1.49	94.22	3.00
E2	98.7	0.14	<1	1.66	3.51	0.93	25.10	2.29
E3	86.0	0.08	<1	1.00	3.38	0.74	18.23	2.62
E4	51.3	0.04	<1	1.20	2.59	0.52	21.29	1.91
E5	48.7	0.05	<1	0.64	3.48	0.21	15.42	1.72
E6	48.7	0.05	<1	1.72	4.04	0.32	39.99	2.22
RR	47.8%	70.6%	/	24.2%	6.3%	78.5%	57.6%	26.0%
W1	83.3	0.14	<1	1.11	2.90	0.99	41.44	2.01
W2	52.7	0.15	<1	2.18	5.86	1.14	39.16	4.07
W3	82.0	0.12	<1	2.19	3.98	1.13	19.47	2.28
W4	86.0	0.08	<1	2.13	5.26	0.63	18.94	2.04
W5	28.7	0.07	<1	2.39	4.67	0.48	21.37	2.67
W6	52.0	0.08	<1	2.37	2.87	0.33	16.73	1.86
RR	37.6%	42.9%	/	/	1.0%	66.7%	59.6%	7.5%
Limit	NA	0.1	≤1	≤10	≤10	20	≤50	≤10

Note: *RR*: removal rate; NA: not available; italic data are above the limits of basic indexes for grade I or supplementary indexes for centralized domestic drinking water sources according to Chinese environmental quality standards for surface water (GB 3838–2002).

Wang *et al.*

Concentrations of heavy metals in all water samples were within the limits of basic indices for grade I according to Chinese environmental quality standards for surface water (GB 3838–2002), except in the case of Zn: Zn content at the eastern inlet (94.22 μg/L) exceeded the threshold (50 μg/L). Thus, ecological risk of heavy metals in the pilot wetland was rather small. The removal rates of Cu, Ni, Zn, and Pb were good, reaching 6.26%, 78.52%, 57.56%, and 26.00% in the east and 1.03%, 66.67%, 59.63%, and 7.46% in the west, respectively. The heavy metal Cr exhibited a good removal rate (24.43%) in the east but a lower rate in the west, although the rate in the west increased slightly with distance along the hydraulic flow pathways. However, these removal rates are less favorable than those of traditional wetlands (Bulc and Slak 2003; Tam et al. 2009), because nearly all heavy metal concentrations were within the standard prescribed limits. Nevertheless, it is clear that this novel type of constructed root-channel wetland has achieved good removal of trace metals in its initial period. Accordingly, such wetlands should be efficient in the mitigation of most heavy metal risks in source water.

Contribution of Aquatic Plants to the Uptake and Accumulation of Metals in Wetland

Nine plant samples of six species of emerged aquatic plants were collected in the Changshuitang pilot wetland. Analysis of these plant samples indicated that aquatic plants had pronounced effects on the uptake and accumulation of metals (Table 6). The samples were collected without destroying plant rhizomes and roots, such that only the aboveground parts of the plant were collected. Compared with recent studies (Wu et al. 2013; Lesage et al. 2007; Bernard and Lauve1995), the tissues of plants in the pilot wetland exhibited moderate heavy metal levels, with high contents of Cr, Cu, Ni, Zn, and Pb. We found reeds (*Phragmites australis*), the dominant wetland species, to have more pronounced effects than other plants on the uptake and accumulation of heavy metals, proving that plant uptake can contribute considerably to reducing heavy metals in the plant bed soils and partly explaining the relatively low heavy metal contents observed in the rhizospheric soil.

Table 6: Heavy metals contents in above ground portion of plants in pilot wetland and other wetlands (mg/kg)

Country	Wetland type	Plant species	Part	Cd	Cr	Cu	Ni	Zn	Pb	Reference
China	Root-channel wetland	Schoenoplectus tabernaemontani	AB	0.04	2.71	7.42	1.72	3.74	0.27	This study
		Typha orientalis	AB	0.03	1.70	2.81	0.70	0.91	0.58	This study
		Typha orientalis	RO	0.03	1.96	0.52	4.15	8.80	1.21	This study
		Juncus effusus	AB	0.17	2.58	6.21	1.21	13.77	1.30	This study
		Acorus calamus	AB	0.16	1.95	5.20	0.62	1.55	0.68	This study
		Phragmites australis	L	0.02	3.19	4.69	1.46	5.27	1.64	This study
		Phragmites australis	St	0.003	2.05	3.87	0.55	15.00	0.45	This study
		Zizania caduciflora	L	0.02	2.59	1.38	1.12	7.54	0.69	This study
		Zizania caduciflora	St	0.02	2.64	2.79	0.49	16.22	0.32	This study
China	Natural wetland	Phragmites australis	Sh	0.12-0.16	2.8-6.4	1.5-1.7	2.7-3.2	10.0-10.2	0.13-0.33	(Wu et al.2013)
Belgium	Constructed wetland	Phragmites australis	St	0.031-0.083	0.9-1.3	2.2-4.4	0.20-0.52	11-39	0.34-0.43	(Lesage et al.2007)
Belgium	Constructed wetland	Phragmites australis	L	0.028-0.071	0.72-0.95	0.91-3.3	0.29-0.45	20-29	0.53-1.1	(Lesage et al.2007)
U.S.A.	Constructed wetland	Phalaris arundinacea	AB			1.9-7.1		10-28		(Bernard and Lauve 1995)

Note: AB: aboveground, L: leaves, St: stems, Sh: shoot, RO: reproduction organ.

Wang *et al.*

DISCUSSION

Metal Accumulation Effects in Zones of CRCW

Constructed root channel wetland (CRCW) can remove many types of pollutants and plays an important role in water pollution control and the ecological rehabilitation of water environments (Wang et al. 2014). However, metals cannot be decomposed; rather, they must be changed morphologically to allow transportation and transformation of their special characteristics (González-Alcaraz et al. 2013; Yeh et al. 2009). When source water first enters the pilot wetland (i.e., in the pretreatment zone), the relatively wide water surface and slow hydraulic flow rates induce primarily the precipitation of TSS and organic matter. Owing to the special configuration of high/low ditches and plant-bed systems, metals can become further enriched in ditches and/or plant beds in root-channel zone I. In the present study, we found metal contents to be relatively low in these previous units, with little deposition in the water lifting and falling zone. The contamination factors of metals in root-channel zone II increased sharply with respect to those in previous zones. Metal deposition in the deep purification zone was much poorer than that in the root-channel zone, although the deep purification zone still promoted effective accumulation of metals owing to its slower hydraulic flow and wider water area. Previous studies have shown that, in constructed wetland, heavy metal contents in sediments typically decline gradually along hydraulic flow pathways (Obarska-Pempkowiak and Klimkowska 1999). Wetlands are believed to select their most favorable evolution direction in accordance with the surrounding environment. In the present study, metal content, contamination factors and *RI* were found to vary dramatically between five functioning zones, particularly between the two root-channel zones. Specifically, *RI*was found to first decline and then increase along the water flow pathways, reaching extremely high values in the two root-channel zones in the second part. This definitively proves that root-channel zones may become "hot sites" acting as sinks for heavy metals in this pilot wetland (Wang et al. 2014). Moreover, the results suggest that obvious deposition and accumulation of heavy metals occurred in CRCW, further highlighting the efficiency of this innovative technology for the mitigation of the ecological risk of metals.

Constructed root-channel wetlands are of particular importance because they can accumulate metals based on their specific characteristics. The retention of metals in the pilot wetland can be attributed primarily to sedimentation effects. Metals are likely to be combined with particulates in water after entering wetland areas, such that sediments are primary sinks for heavy metals (Zhao et al. 2012; Accornero et al. 2008; Sajwan et al. 2008). Also, the dominated aquatic plant reeds (*Phragmites australis*) played important role in sequestering metals (Weis and Weis 2004). In root channel zones, the rhizospheric microbial characteristics affected the metal accumulation (Wu et al. 2013). Phytoaccumulation occurs when heavy metals are not degraded in the plant, resulting in its accumulation within the plant tissues and the potential for plant uptake is highest in CWs due to the increased contact between water and the elaborate root systems of aquatic macrophytes (Malaviya and Singh 2012). In Tables 7 and 8, fluxes of heavy metals in sediment/soil, sedimentation and plant uptake masses of heavy metals in pilot wetland were calculated. Sedimentary mass of heavy metals accounted for 63.30% and soil part was 36.67%. However, percentage of metal mass by plant uptake was less than 0.1%. The pilot wetland can maintain about 86.18 kg total heavy metals since its operation. The fluxes for sediment ranged from 0.41-211.08 $\mu g \cdot cm^{-2} \cdot a^{-1}$, less than that in soils (0.73-543.94 $\mu g \cdot cm^{-2} \cdot a^{-1}$), both less than those in Shijiuyang Wetland (Wang et al. 2014).

Table 7: Masses of heavy metals accumulated in pond/ditch sediments, plant bed soils and plants

Structural unit	Type	Area/m2	Cd/g	Cr/g	Cu/g	Ni/g	Zn/g	Pb/g	Sum/g	Percentage
Pond	Sediment	3615.43	20.19	9671.14	2412.70	3948.40	7237.27	3408.46	26698.15	30.98%
Ditch	Sediment	3554.16	20.68	10153.41	2370.52	3995.47	8133.73	3182.92	27856.72	32.32%
Plant bed	Soil	1740.14	17.26	12810.70	2917.46	5144.17	7878.62	2832.86	31601.07	36.67%
First part	Plant	735.40	0.07	1.65	3.20	0.97	5.58	0.72	12.19	0.01%
Second part	Plant	1004.74	0.03	2.46	3.89	1.50	7.70	0.68	16.26	0.02%
Total			58.22	32639.36	7707.76	13090.50	23262.91	9425.64	86184.39	100.00%

Wang *et al.*

Table 8: Accumulation fluxes of heavy metals in pond/ditch sediments and plant bed soil ($\mu g \cdot cm^{-2} \cdot a^{-1}$)

Structural unit	Type	Operational days	Cd	Cr	Cu	Ni	Zn	Pb
Pond	Sediment	494	0.41	197.64	49.31	80.69	147.90	69.66
Ditch	Sediment	494	0.43	211.08	49.28	83.06	169.09	66.17
Plant bed	Soil	494	0.73	543.94	123.88	218.42	334.53	120.28

Wang *et al.*

The east and west blocks of the wetland described here were designed with almost identical theoretical operating conditions, with similar characteristics in terms of construction, operation, and local plant conditions, yet they resulted in varied soil/sedimentary metal contents. However, despite the almost symmetrical and parallel design of the east and west blocks, the results presented here indicate non-parallel enrichments effects between the two blocks. This can be attributed to slight differences in local conditions, development, succession, human activities, and hydraulic factors (Stead-Dexter and Ward 2004; Wu et al. 2013). However, a more full characterization of these differences and their effects will require long-term comparative study, which will help to improve understanding of the system and allow optimization of engineering design and operation and management practices, thus allowing high purification efficiencies to be achieved consistently.

Metals Accumulation Effects During Operating Periods of CRCW

The Changshuitang wetland is located within a dense stream network. Accordingly, it is subjected to extremely complex water sources, various pollutant types, and serious nonpoint and diffuse but small decentralized point source pollution (Yin et al. 2010). Pollutants such as heavy metals enter the water body of stream networks through overland runoff in cities. Previous studies have shown that the first full-scale CRCW with a similar structural design to that of the present study, the Shijiuyang wetland, demonstrated high enrichment factors for Cd, Cr, Cu, Ni, Zn, and Pb; these values remained above the threshold value of 1.5 after 4 years of operation (Wang et al. 2014). Both studies showed that the crisscrossed plant-bed/ditch systems in the root-channel zones were "hot sites" for the sedimentation, interception, enrichment, and even uptake of heavy metals in CRCW. Moreover, the risks of heavy metals were greatly mitigated through the complex interactions of the plant-bed/ditch systems. The removal rates of heavy metals in the water phase correspond well to the characteristics of the sediment and soil matrix (particularly particle size composition and organic matter content), and the interactions among plant-bed soil matrixes and the bilateral ditch water are capable of removing most species of metals, in conjunction with the deposition and filtration of coarse/finer particles.

Although the Changshuitang pilot wetland has been operational for only 1 year and performed poorly compared with the Shijiuyang wetland in terms of plant diversity and metal removal rate, it achieved better water quality using a more reasonable design for similar environmental and soil matrix conditions. In particular, its unique design helped enhance removal rates for metals and heavy metals. However, no extremely obvious decline of *RI* was observed in the pilot wetland, particularly compared with that between the wetland inlet (Site A, source water sediment) and the wetland outlet (Site G, exit of deep purification zone). This may be due to instability of wetland structure and function, low biodiversity, and the unstable effects of water purification (Saeedreza et al. 2012). The effects of metal accumulation in the pilot wetland's primary operation (one-year period) were comparatively lower than those in the relatively developed Shijiuyang wetland (four-year period). In addition, it may be important that the hydraulic flow pathways (direct length ca. 200–300 m) and hydraulic retention time (ca. 2.3 d) are both relatively small under a hydraulic flow rate of 0.27 m/d. This may be insufficient for thorough interaction between water and sediments/soil (Bilal et al. 2009). Moreover, the very low concentrations of metals (except Zn) in the source river water do not provide a considerable pollution load to the wetland, where the optimum treatment efficiency of such metals may be achieved under conditions in which an allochthonous source input is present. At present, *RI* in the pilot wetland is far below the threshold value of 150, indicating little environmental damage. Nevertheless, metal contents at the outlet of the pilot wetland remain relatively high owing to the effects of accumulation; this can be attributed to the internal interweaved "capillary" filtration functions, which can be visualized by considering the whole pilot wetland as the glomerulus of a kidney.

Decreases in flow velocity at the restricted exit of the wetland will promote metal deposition accompanied by particulate settling (Bilal et al.2009; Brix and Arias 2005). Nonetheless, more long-term monitoring data will be required to assess the actual removal of metals and the transformation of metal forms. Moreover, further emphasis should be placed on the effects of long-term operation, accumulation effects, the use of aquatic plants, and the effects of novel CRCW techniques on the removal of metals (especially heavy metals) in future.

CONCLUSIONS

The metals exhibited considerable spatial variation within the pilot wetland. According to the generally adopted sediment quality assessment values for freshwater ecosystems, the pilot wetland showed rare or occasional adverse effects by heavy metals. Most sample sites in the wetland had contamination factors above one and the maximum contamination factors for heavy metals were all above one except Cd in west wetland and Cu in east wetland. As contamination factors showed, the plant-bed/ditch system primarily in the second part of the Changshuitang pilot wetland accumulated heavy metals (Cd, Cr, Cu, Ni, Zn, and Pb). The comprehensive potential ecological risk index (*RI*) decreased gradually along the hydraulic flow pathways in the first part and increased in the second part indicating that both the pre-pond and the post-pond functioned effectively to retain metals. Small ditches within the plant-bed/ditch systems in the root-channel zone demonstrated the ability to remove metals from source water. *RI* exhibited spatial variation, decreasing in the following order: high ditch, low ditch, plant bed. Heavy metals in the source water became considerably accumulated and enriched within the plant-bed/ditch systems in the root-channel zones. The wetland effectively eliminated Cu, Ni, Zn, and Pb, with high removal rates. Sedimentary mass of heavy metals accounted for 63.30% and soil part was 36.67% in this wetland. However, percentage of metal mass by plant uptake was less than 0.1%. The fluxes for sediment ranged from 0.41-211.08 $\mu g \cdot cm^{-2} \cdot a^{-1}$, less than that in soils (0.73-543.94 $\mu g \cdot cm^{-2} \cdot a^{-1}$). The pilot wetland can maintain about 86.18 kg total heavy metals since its operation. This pilot case study proves that a purification mechanism formed by a combination of multilevel ponds and plant-bed/ditch systems can retain heavy metals from the micro-polluted source water, even during its first year of operation, thus reducing the potential ecological risk of the purified raw water. The pre-pond, plant-bed/ditch, and post-pond complexes can provide a new and effective method for the removal of heavy metals in drinking water sources.

MATERIALS AND METHODS

Background and Study Area

The water source in the river networks of the Yangtze River delta of China has become increasingly polluted in response to rapid economic development and urbanization within the watershed. Jiaxing belongs to the delta with a population of 4.5 million, where heavy metal is one of the potentially important environmental issues. Many industrialized processes including dying, plating, and tanning give rise to the contamination by heavy metals, such as cadmium (Cd), mercury (Hg) and lead (Pb) in soil, water and air (Huang et al. 2013). Busy land and water transportation in the Great Canal are also an important input source for the stream network. Urban runoff caused by rainfall on city dusts is one of the primary pathways of heavy metal into the stream network, and the narrow land/water boundaries, absence of buffer zones, and ineffective street cleaning methods contribute to the heavy metal inputs to the stream network (Zhao et al.2009). Jiaxing has no mountain reservoir for water source and has to take drinking water from river network. Although considerable efforts have been made to control the pollution in the city's stream network, some of the drinking water in the network remains polluted owing to the complex conditions in the study area, thus threatening water safety in the region (Yin et al. 2010).

On April 1, 2007, Jiaxing began to construct the Shijiuyang large-scale ecological water quality treatment wetland (Figure 1), which was intended to purify and restore the healthy drinking source water. This treatment system was based on three patented technologies. The core technology focused on constructing root-channels artificially in constructed wetland and was developed by the Research Center for Eco-Environmental Sciences (RCEES), Chinese Academy of Sciences

(CAS) (Wang et al. 2012a). The trial operation of the wetland was initiated on July 2, 2008. This full-scale wetland incorporated root-channel renewability to prevent clogging. To date, the wetland has been operating continuously for nearly 6 years realizing the goal of improving source water quality by one level according to Chinese environmental quality standards for surface water (GB 3838–2002). Furthermore, the wetland has become a city park serving multiple ecological functions. This innovative technique put into practice by Jiaxing both guarantees drinking water safety for the local population and promotes continuous improvement of the ecological and living environments. Moreover, it provides a meaningful example for other cities that may be facing similar problems relating to micro-pollution of their drinking water.

Thanks to the success of the Shijiuyang wetland, the Changshuitang wetland was constructed in Haining, becoming the third CRCW in China (Figure 1). The Changshuitang wetland benefited from many improvements over the Shijiuyang wetland. For example, it benefited from the design and application of many optimization measures, including increased diversity of pioneer plants, application of two-way parallel water flow pathways, strengthening of hydraulic control measures, and adoption of a step-feed process of water supply to enhance ammonia and organic matter removal. These measures were adopted primarily to further mitigate the complex pollution arising from industrial, domestic, and agricultural sources in the drinking water sources represented by the Changshanhe and Changshuitang rivers, which are treated by the Third Water Plant in Haining. The Changshuitang wetland is located in the northern part of Haining, which lies to the east of the Third Water Plant. The wetland covers an area of about 1.73 km^2 and aims to purify about 0.3 million m^3 of drinking water per day. The wetland itself was designed to use the natural topography of the original course of the river, based on the concept of ecological treatment, and was constructed according to the patent technologies of RCEES, as described above. Today, the wetland is still under construction.

To further optimize, promote, and verify the water treatment efficiency of the full-scale wetland (ca. 1.73 km^2), a pilot wetland at a scale of approximately 1:100 (ca. 1.83 ha) was built beforehand nearby from March to June 2011. Its geographical coordinates span 30°34′5.36″–30°34′11.82″ N and 120°42′23.32″–120°42′27.79″ E. It was designed as two parallel wetland blocks divided by a central water supply channel that serves as a partial water source (i.e., accounting for 20 - 50% of the total water supply amount) to allow step feeding of the wetland through the pump station and through horizontal water pipes in the first and second parts, respectively (Figure 1). This configuration results in east and west wetlands for separate purification of the micro-polluted drinking water source. Each block (wetland) is composed of five functioning zones, which are further divided into two parts by a pump station. The first part is located before the pump station and contains pretreatment zone and root-channel zone I, whereas the second part lies after the pump station and contains the water lifting and falling zone, root-channel zone II, and the deep purification (polishing) zone. Each root-channel zone consists of finer scale structural units including a high water level ditch (high ditch), low water level ditch (low ditch), and plant bed. The pilot wetland can deal with 1880 m^3 water per day and typically improves water quality by one level.

Sample Collection and Preparation

In total, 71 sediment samples from ponds/ditches and 16 soil samples from plant beds were collected on December 7–8, 2012 (Figure 1). We assigned sample sites in representative areas of the wetland to obtain information regarding key processes. As the area of pilot wetland was small, boats could not be used to collect samples. Then sample sites in water lifting and falling zone as well as deep purification zone were not in the middle part but we try our best to collect the most representative samples. The sampling site numbers were assigned as follows. Single "C" before site names indicates sites in the central channel, with C1 for the inlet and C2 in sites beside the water pump stations controlling the step-feed process. Similarly, the use of prefix "W" and "E" before site names indicate sites in the west and east blocks, respectively. Two equivalent sedimentary samples from the inlets of the east and west sides were mixed as one; this sample is labeled "Source" (or "A"). According to the hydraulic flow pathways, the first part lies before the pump station and contains pretreatment zone (B) and root-channel zone I (C), whereas the second part includes the water lifting and falling zone (D), root-channel zone II (E), and the deep purification zone (F). The outlets of the wetlands are indicated by "Outlet" (or "G"). Within each zone, we used numbers to label different sites along the water flow pathways. In each root-channel zone, the ditches lead the water deep into the wetland and let the water penetrate through the root-channels under the plant beds. "H" and "L" refer to ditches with high and low water levels, respectively, and "P" indicates sites of soil samples from plant beds.

About 400–500 g wet weight of mixed surface sediments (0–10 cm) were collected using an Ekman bottom sampler (Hydro-Bios, Kiel, Germany). To reduce the heterogeneity of sediments and avoid the effects of water and plant, three equal samples were collected and mixed as one. Soil samples from plant beds weighing 300–400 g were collected using a soil auger (XDB-TR7, sampling depth: 1.5 m, drill diameter: 5 cm; Beijing New Landmark Soil Equipment Co., Ltd.) at depths of 0–30 cm. Immediately after collection, soil temperature, pH, and redox potential (E_h) were measured using an IQ150 probe (HACH, USA); then, samples were preserved in bags on ice for transportation. Samples for ammonium-N, nitrate-N, and nitrite-N assays were stored under -4 °C. The remaining samples were frozen, air dried, and sieved through a 100-mesh sieve to remove coarse particles. The sediment/soil samples were then stored in plastic bags for further analysis.

Twelve water and nine plant samples were collected on August 3, 2013 for further investigation of metal contents and risk levels in the water column and plants of the wetland in full flourish. Water samples are indicated by "W" (west wetland) or "E" (east wetland) and numbers, while plant samples are expressed by "P" and numbers. Samples W1–6 and E1–6 represent key sites at the intersections between two functioning zones, i.e., the inlet of the wetland (where the source water feeding the wetland enters the system), outlet of the pretreatment zone, outlet of root-channel zone I, outlet of the water lifting and falling zone, outlet of root-channel zone II, and outlet of the deep purification zone (i.e., the exit of the whole wetland). Water samples for the analysis of metal contents were stored after filtering through 0.45-μm filter membranes before adding HNO_3. General water samples were stored at cold temperatures after their temperature, pH, dissolved oxygen (DO), and redox potential (E_h) had been analyzed. All types of emerged aquatic plants present were sampled to provide a representative indication of plant conditions in the wetland; such plant samples were stored in plastic bags after determination of their stem diameter, height, and density from random samples.

Analysis and Quality Control

For metal extraction, 0.1-g samples of dried soil were digested in 6 mL of an HNO_3 and HCl mixture (1:3) and 2 mL HF in a CEM microwave (MARS Xpress; CEM, USA) according to the program recommended by the United States Environmental Protection Agency (EPA) (Bettinelli et al. 2000). These transparent solutions were then filtered through 0.45-μm filter membranes and diluted to 50 mL with distilled water. The concentrations of metals Cd, Cr, Cu, Ni, Pb, and Zn in the filtrate were determined by ICP–MS (7500a; Agilent, USA), whereas the concentrations of metals K, Na, Ca, Mg, Al, and Fe were measured by ICP–OES (OPTIMA 2000DV; Perkin Elmer, USA).

Determination of metal content of water samples was achieved by ICP-OES (OPTIMA 2000DV; Perkin Elmer) for K, Na, Ca, Mg, and Fe and by ICP-MS (7500 a; Agilent) for Al, Cd, Cr, Cu, Ni, Zn, and Pb. Total suspended solids (TSS) were measured by mass according to standard methods. Plant samples were digested by HNO_3-H_2O_2 in a CEM after drying and sieving; further details are provided in the existing literature (Niemelä et al. 2004). The subsequent metal determination methods for plant samples were the same as those for soil/sediment samples.

The metals can be classified into groups/types according to several classification systems in the context of study objectives. To ensure clarity, we refer to the metals K, Na, Ca, Mg, Al, and Fe as major metals owing to their abundance in the study area (their contents are given in %) and refer to the metals Cd, Cr, Cu, Ni, Zn, and Pb (i.e., those measured in mg/kg) as heavy metals, following the most commonly used terminology. Major metals were not the wide discussion in this study with only necessary description.

Enrichment Factor

The enrichment factor (*EF*) is typically used to evaluate sources of metals and requires the use of a standard element that fulfills several criteria (Abrahim and Parker 2008). We adopted Fe as the standard element (Cobelo-García and Prego 2003) and calculated *EF* as follows.

$$EF = \frac{(Me / Fe)_{Sample}}{(Me / Fe)_{background}} \tag{1}$$

Here, $(Me/Fe)_{sample}$ and $(Me/Fe)_{background}$ refer to the ratios of a target metal to that of Fe in a soil/sediment sample and in the background, respectively. We adopted metal contents obtained for the surface soil in the Hangzhou–Jiaxing–Huzhou plain (soil layer A) as the background content (Wang et al. 2007). Higher values of *EF* indicate more extensive accumulation of metals. $EF < 1.5$ signifies that metal contents are natural levels; thus, $EF > 1.5$ suggests that metals resulting from human activities are a key component (Håkanson 1980).

Assessment of Potential Ecological Risk Index of Heavy Metals

The potential ecological risk index (*RI*) of heavy metals can be used to assess the ecological risk they pose. Such a risk is typically assessed using an index that reflects the content of heavy metals, the number of heavy metal pollutant sources, the toxicity level, and any ecological/environmental effects, as follows (Håkanson 1980).

$$RI = \sum E_y^i = \sum T_y^i \times C_y^i = \sum T_y^i \times \frac{C_s^i}{C_n^i} \tag{2}$$

Here, C_s^i and C_n^i represent the content of heavy metal *i* and its background content in Hangzhou–Jiaxing–Huzhou plain (soil layer A), respectively. Thus, C_f^i (i.e., $\frac{C_s^i}{C_n^i}$) represents the contamination factor of heavy metal *i*. T_y^i indicates the toxic response factor for a given heavy metal, where T_r^i is 30, 2, 5, 5, 1, and 5 for Cd, Cr, Cu, Ni, Zn, and Pb, respectively (Håkanson 1980), E_r^i represents the potential ecological risk index of an individual heavy metal, and *RI* is the sum of the potential risk of all individual heavy metals.

Fluxes and Masses of Heavy Metals

Flux of heavy metals in sediment/soil can be calculated with indices such as content of metal, moisture content and dry bulk density (Wang et al. 2014). It represents the rate of sedimentary effects on heavy metals.

$$F_i = C_i \times (1 - W) \times p \times \Delta H / t \quad (3)$$

Here, *Fi* means fluxes of heavy metal *i* in sediment/soil ($mg \cdot m^{-2} \cdot d^{-1}$), *Ci* means contents of metal *i*($mg \cdot kg^{-1}$); *W* indicates moisture content; ρ indicates average dry bulk density ($kg \cdot m^{-3}$); ΔHindicates thickness of sediment/soil (m); *t* represents operational days (d).

Mass of heavy metals accumulated or absorbed by sediment/soil and plants is calculated by formula 4 and 5.

$$M_i = F_i \times S_i \times t \quad (4)$$

Here, *Mi* indicates accumulation mass of metal *i* (mg); *Sj* means area of zone *j* (m^2).

$$M_p^i = C_p^i \times B_j \times S_p^j \times n \quad (5)$$

Here, M_p^i indicates uptake mass of heavy metal *i* by plants (mg); C_p^i means content of metal *i* in plants ($mg \cdot kg^{-1}$); *Bj* indicates biomass of plants ($kg \cdot m^{-2}$); S_p^i means area of aquatic plants (m^2); *n*represents times of reaping.

Statistical Analysis

Data analysis was performed using SAS for Windows 9.2 (SAS Institute, Inc., Cary, NC, USA) (Friendly 1991; Friendly 2000). Unless otherwise stated, $\alpha = 0.05$ and $\alpha = 0.01$ were adopted as the statistically significant and extremely significant levels. A biplot was drawn using the IML module in the SAS system to visualize the multivariate relationships between observations and variables. In this plot, the lengths of environmental vectors represent the ability to distinguish different variables and the cosine of angles between two vectors expresses the degree of correlation. All variables in this biplot were standardized to ensure consistency in units and dimensions.

AUTHORS' CONTRIBUTIONS

BW carried out the whole experimental studies, drafted and revised the manuscript. YW participated in the field research and indoor experiments, and helped to analyze the experimental results. WW designed the pilot system and conceived of the study, performed the statistical analysis and drawing, and helped to revise the manuscript. BW also involved in data analysis and interpretation. All authors read and approved the final manuscript.

ACKNOWLEDGEMENTS

This research was supported by the National Natural Science Foundation of China (51278487), the Knowledge Innovation Program of the Chinese Academy of Sciences (KZCX2-EW-410-05), and the Major National Water Pollution Control and Management Project (2012ZX07403-003-03). The authors thank Professor Chengqing Yin and Dr. Guibing Zhu for contribution to the pilot design, and thank Ms. Xiaohong Wei and Jinzhu Yang for final polishing of the pilot design. Special thanks are given to Mr. Yicheng Shen for his great help in field sampling and transportation. We are grateful to Professor Zhimin Qiang for his constructive suggestions of revamping the manuscript. The authors greatly appreciate the editor Dr. Lynsey Callaghan who worked on behalf of Liwen Bianji (Edanz Group China) for editorial comments and language corrections.

REFERENCES

1. Abrahim GMS, Parker RJ (2008) Assessment of heavy metal enrichment factors and the degree of contamination in marine

sediments from Tamaki Estuary Auckland New Zealand. Environ Monit Assess 136(1–3):227-238

2. Accornero A, Gnerre R, Manfra L (2008) Sediment concentrations of trace metals in the Berre lagoon (France): an assessment of contamination. Arch Environ Contam Toxicol 54(3):372-385
3. Bernard JM, Lauve TE (1995) A comparison of growth and nutrient uptake in *Phalaris arundinacea*L growing in a wetland and a constructed bed receiving landfill leachate. Wetlands 15(2):176-182
4. Bettinelli M, Beoneet GM, Spezia S (2000) Determination of heavy metals in soils and sediments by microwave-assisted digestion and inductively coupled plasma optical emission spectrometry analysis. Anal Chim Acta 424(2):289-296
5. Bilal T, Selma A, Lütfi A, Kemal G (2009) Performance of a pilot-scale three-stage constructed wetland system for domestic wastewater treatment. Environ Technol 30(11):1187-1194
6. Brix H, Arias CA (2005) Danish guidelines for small-scale constructed wetland systems for onsite treatment of domestic sewage. Water Sci Technol 51(9):1-9
7. Bulc T, Slak AS (2003) Performance of constructed wetland for highway runoff treatment. Water Sci Technol 48(2):315-322
8. Cobelo-García A, Prego R (2003) Heavy metal sedimentary record in a Galician Ria (NW Spain): background values and recent contamination. Mar Pollut Bull 46(10):1253-1262
9. Friendly M (1991) SAS system for statistical graphics. SAS Institute Inc, Cary, NC, USA. pp 437-445, 532-536
10. Friendly M (2000) Visualizing categorical data. SAS Institute Inc., Cary, NC, USA. pp 339-341
11. González-Alcaraz MN, Miguel Conesa H, Álvarez-Rogel J (2013) Phytomanagement of strongly acidic saline eutrophic wetlands polluted by mine wastes: the influence of liming and *Sarcocornia fruticosa* on metals mobility. Chemosphere 90(10):2512-2519
12. Goulet RR, Leclair EN, Pick FR (2001) The evaluation of metal retention by a constructed wetland using the pulmonate gastropod *Helisoma trivolvis* (Say). Arch Environ Contam Toxicol 40(3):303-310

13. Graney JR, Eeriksen TM (2004) Metals in pond sediments as archives of anthropogenic activities: a study in response to health concerns. Appl Geochem 19(7):1177-1188
14. Håkanson L (1980) An ecological risk index for aquatic pollution control: a sedimentological approach. Water Res 14(8):975-1001
15. Huang Z, Pan XD, Wu PG, Han JL, Chen Q (2013) Health risk assessment of heavy metals in rice to the population in Zhejiang, China. PLoS One 8(9):e75007 doi:10.1371/journal.pone.0075007
16. Kadlec RH (2003) Pond and wetland treatment. Water Sci Technol 48(5):1-8
17. Lesage E, Rousseau DP, Meers E, Tack FM, De Pauw N (2007) Accumulation of metals in a horizontal subsurface flow constructed wetland treating domestic wastewater in Flanders Belgium. Sci Total Environ 380(1):102-115
18. MacDonald DD, Ingersoll CG, Berger TA (2000) Development and evaluation of consensus-based sediment quality guidelines for freshwater ecosystems. Arch Environ Contam Toxicol 39(1):20-31
19. Malaviya P, Singh A (2012) A Constructed wetlands for management of urban stormwater runoff. Crit Rev Environ Sci Technol 42(20):2153-2214
20. Mohammed I, Abdu N (2013) Horizontal and vertical distribution of lead, cadmium, and zinc in farmlands around a lead-contaminated goldmine in Zamfara, Northern Nigeria. Arch Environ Contam Toxicol 66(2):1-8
21. Niemelä M, Perämäki P, Piispanen J, Poikolainen J (2004) Determination of platinum and rhodium in dust and plant samples using microwave-assisted sample digestion and ICP-MS. Anal Chim Acta 521(2):137-142
22. Obarska-Pempkowiak H, Klimkowska K (1999) Distribution of nutrients and heavy metals in a constructed wetland system. Chemosphere 39(2):303-312
23. Protano C, Guidotti M, Owczarek M, Fantozzi L, Blasi G, Vitali M (2013) Polycyclic aromatic hydrocarbons and metals in transplanted Lichen (*Pseudovernia furfuracea*) at sites adjacent

to a solid-waste landfill in central Italy. Arch Environ Contam Toxicol 66(4):1-11

24. Rai PK (2012) An eco-sustainable green approach for heavy metals management: two case studies of developing industrial region. Environ Monit Assess 184(1):421-448
25. Rana S, Bag SK, Jana BB, Biswas JK (2013) Seasonal distribution of cadmium among components of sewage treatment ponds: an eco-tech for heavy metal remediation. Int J Environ Sci Tech 10(5):1103-1114
26. Saeedreza H, Jin-Lee KM, Jeremy R (2012) Evaluating removal efficiency of heavy metals in constructed wetlands. J Environ Eng 138(4):475-482
27. Sajwan KS, Kumar KS, Paramasivam S, Compton SS, Richardson JP (2008) Elemental status in sediment and American oyster collected from Savannah marsh/estuarine ecosystem: A preliminary assessment. Arch Environ Contam Toxicol 54(2):245-258
28. Si W, Ji W, Yang F, Lv Y, Wang Y, Zhang Y (2011) The function of constructed wetland in reducing the risk of heavy metals on human health. Environ Monit Assess 181(1–4):531-537
29. Stead-Dexter K, Ward NI (2004) Mobility of heavy metals within freshwater sediments affected by motorway stormwater. Sci Total Environ 334:271-277
30. Tam NFY, Wong YS, Wong MH (2009) Novel technology in pollutant removal at source and bioremediation. Ocean Coast Manage 52(7):368-373
31. Vymazal J (2011) Constructed wetlands for wastewater treatment: five decades of experience. Environ Sci Technol 45(1):61-69
32. Vymazal J, Krása P (2003) Distribution of Mn Al Cu and Zn in a constructed wetland receiving municipal sewage. Water Sci Technol 48(5):299-305
33. Wang QH, Dong YX, Zhou GH, Zheng W (2007) Soil geochemical baseline and environmental background values of agricultural regions in Zhejiang Province. J Ecol Rural Environ 23(2):81-88 (in Chinese)
34. Wang ZQ, Wang WD, Zhu GB, Yin CQ (2011) A comparative study on the diversity of rhizospheric bacteria community structure in

constructed wetland and natural wetland with reed domination. Acta Ecologica Sinica 31(16):4489-4498 (in Chinese)

35. Wang WD, Wang ZQ, Li J, Yin CQ (2012) The constructed root-channel technology and its application in constructed wetlands. Acta Sci Circum 32(1):43-50 (in Chinese)
36. Wang ZQ, Zhang RB, Chen QH, Wei HB, Wang WD (2012) Water treatment efficiency of constructed wetland plant-bed/ditch systems. Environ Sci 33(11):3804-3811 (in Chinese)
37. Wang Y, Wang BL, Wang WD (2014) Sources distribution and accumulation effect of metal elements in a constructed root-channel wetland. Acta Sci Circum 34(1):168-185 (in Chinese)
38. Weis JS, Weis P (2004) Metal uptake transport and release by wetland plants: implications for phytoremediation and restoration. Environ Int 30(5):685-700
39. Wu JT, Wang L, Ma F, Yang JX, Li SY, Li Z (2013) Effects of vegetative-periodic-induced rhizosphere variation on the uptake and translocation of metals in *Phragmites australis* (Cav.) Trin ex. Steudel growing in the Sun Island Wetland. Ecotoxicology 22(4):608-618
40. Xiao R, Bai JH, Gao HF, Wang JJ, Huang LB, Liu PP (2012) Distribution and contamination assessment of heavy metals in water and soils from the college town in the Pearl River Delta China. CLEAN-Soil Air Water 40(10):1167-1173
41. Yang HJ, Shen ZM, Zhu SH, Wang WH (2008) Heavy metals in wetland plants and soil of Lake Taihu China. Environ Toxicol Chem 27(1):38-42
42. Yeh TY, Chou CC, Pan CT (2009) Heavy metal removal within pilot-scale constructed wetlands receiving river water contaminated by confined swine operations. Desalination 249(1):368-373
43. Yin CQ, Su SL, Zhang RB, Wang WD, Zhou CD, Liu JX, Zhao HT (2010) Use of wetland to purify polluted stream network for a city water source. Acta Sci Circum 30(8):1583-1586 (in Chinese)
44. Zhang SZ, Wang SX, Shan XQ (2002) Distribution and speciation of heavy metals in surface sediments from Guanting Reservoir Beijing. J Environ Sci Health Pt A 37(4):465-478

45. Zhao HT, Yin CQ, Chen MX, Wang WD (2009) Risk assessment of heavy metals in street dust particles to a stream network. Soil Sediment Contam 18:173-183
46. Zhao Y, Yang ZF, Xia XH, Wang F (2012) A shallow lake remediation regime with *Phragmites australis*: incorporating nutrient removal and water evapotranspiration. Water Res 46(17):5635-5644
47. Zheng J, Chen QH, Zhang RB, Wang WD, Wei HB, Yin CQ (2012) Algae trapping function of plant-bed/ditch systems. Chin J Environ Eng 6(12):4263-4267 (in Chinese)
48. Zheng J, Wang WD, Yin CQ (2012) Distribution and retention of PAHs in a constructed wetland in the Yangtze River delta China. Fresenius Environ Bull 21(9):2594-2602

Chapter 6

Investigation of Heavy Metals in Frequently Utilized Medicinal Plants Collected from Environmentally Diverse Locations of North Western India

Alpana Kulhari[1], Arun Sheorayan[1], Somvir Bajar[2], Susheel Sarkar[3], Ashok Chaudhury[1], and Rajwant K Kalia[4]

of Bio and Nanotechnology, Guru Jambheshwar University of Science and Technology, Hisar 125001, India

[2]Department of Environmental Science and Engineering, Guru Jambheshwar University of Science and Technology, Hisar 125001, India

[3]Indian Agricultural Statistics Research Institute, Library Avenue, Pusa, New Delhi, India

[4]Central Arid Zone Research Institute (CAZRI), Jodhpur, Rajasthan 342003, India

ABSTRACT

The increasing prevalence of environmental pollution, especially soil contamination with heavy metals has led to their uptake in the human food chains through plant parts. Accumulation and magnification of heavy metals in human tissues through consumption of herbal remedies can cause hazardous impacts on health. Therefore, chemical profiling of nine heavy metals (Mn, Cr, Pb, Fe, Cd, Co, Zn, Ni and Hg) was undertaken in stem and leaf samples of ten medicinal plants (*Acacia nilotica, Bacopa monnieri, Commiphora wightii, Ficus religiosa, Glycyrrhiza glabra,Hemidesmus indicus, Salvadora oleoides, Terminalia bellirica, Terminalia chebula* and *Withania somnifera*) collected from environmentally diverse regions of Haryana and Rajasthan states in North-Western India. Concentration of all heavy metals, except Cr, was within permissible limits in the tested stem and leaf samples. Leaf samples had consistently more Cr compared to respective stem samples with highest concentration in leaf samples of *Bacopa monnieri* (13.19±0.0480 ppm) and stem samples of *Withania somnifera* (4.93±0.0185 ppm) both collected from Bahadurgarh (heavy industrial area), Haryana. This amount was beyond the permissible limit of 2.0 ppm defined by WHO for raw herbal material. Other two most perilous metals Pb (2.64±0.0260) and Cd (0.04±0.0274) were also recorded in Bahadurgarh region, although below permissible limits. Concentration of Hg remained below detectable levels in all the leaf and stem samples tested. These results suggested that cultivation of medicinal plants and other dietary herbs should be curtailed near environmentally polluted especially industrial areas for avoidance of health hazards.

INTRODUCTION

Herbal medicines being natural are preferred over synthetic remedies by a major section of the world. Since times immemorial, ancient Indians, Egyptians, Chinese and others have employed a variety of plants and plant products for curing all kinds of ailments. Approximately, 25000 plant based formulations are available in the indigenous medical texts (Gupta et al., 2004). Also, the modern pharmacopoeia contains at least 25% drugs derived from plants and many others which are synthetic analogues built on prototype compounds isolated from plants (Rao et al., 2004). The World Health Organization (WHO) estimated that 80% of the population of developing countries relies on traditional medicines, mostly plant drugs, for their primary health care needs (Planning Commission Report, 2000). Complementary and alternative medicine has received global attention in the last few years with a frequency of 31.4% population in industrialized societies, 42–69% in United States, 71% Canadians and 90% British population consuming dietary supplements or natural health products (vitamins, minerals, amino acids, essential fatty acids, herbal products, traditional Chinese medicines, homeopathic medicines, and probiotics) for treatment, including elimination of disease causing agents, avoidance of side effects and for getting quality life (Szeto,2007). The world market for herbal medicine, including herbal products and raw materials has been estimated to have an annual growth rate between 5 and 15%. Total global herbal drug market is estimated at US $62 billion and is expected to grow to US $5 trillion by the year 2050 (Joshi et al., 2004).

Accumulation of toxic industrial effluents in the soil is continuously increasing due to fast urbanization and extensive pollution of the environment. Among these toxic substances, presence of heavy metals (HMs; atomic weights 63.5–200.6 g mol^{-1} and a specific gravity greater than 5 g cm^{-3}) which are ubiquitous in nature, cause serious harmful effects on living organisms (Nies, 1999; Lee and Lee, 2002). Plants are sensitive to environmental conditions and they accumulate these HMs in their harvestable parts (via root uptake, foliar adsorption and deposition of specific elements in leaves) and intensity of this uptake process changes the overall elemental composition of the plant (Olajire and Ayodele, 2003). Uptake, accumulation and concentration of HMs in plants is influenced by various attributes including atmospheric depositions (depend on traffic densities, metal mining and smelting operations), concentration and bioavailability of HMs in soil (through addition of pesticides and sewage sludge), the nature of soil where herbs are grown (pH and organic matter concentration), individual plant performance (degree of maturity of the plant, time of harvest) and manufacturing conditions of herbal drugs (grinding weights, lead-releasing containers and manufacturing utensils) etc (Nwoko and Mgbeahuruike, 2011). Different HMs have different transmitting rates from soil-to-plant, based on transfer coefficients of metals viz: Cd, Tl and Zn are readily taken up by plants because of higher transfer coefficient, whereas Cu, Co, Cr, and Pb are stably bound to the soil structures and show minimum transfer to plants from soil due to lower transfer coefficient (Kloke et al., 1984). Some metals (Mg, Mn and Zn) play a vital role in proper growth and development of the plant being directly or indirectly involved in various biological functions of enzyme activation and molecular metabolism. Very little information is available about potential influence of metals on pharmacological activity of natural drugs obtained from medicinal plants. Metal mediated hazardous impacts can be direct or indirect via binding of metals with pharmacologically active substances or by manipulating the pharmacokinetics (Weber and Konieczyski, 2003).

This study was designed to investigate the levels of nine heavy metals (Mn, Cr, Pb, Fe, Cd, Co, Zn, Ni and Hg) in leaf and stem samples of ten medicinal plants (*Acacia nilotica, Bacopa monnieri,Commiphora wightii, Ficus religiosa, Glycyrrhiza glabra, Hemidesmus indicus, Salvadora oleoides,Terminalia bellirica, Terminalia chebula* and *Withania somnifera*) collected from environmentally diverse regions of North Western India (Haryana and Rajasthan states) using AAS. The collection sites were shortlisted on the basis of presence of contaminated soil and atmospheric polluting units (presence of industries, heavy traffic sites and site located near to petrol pumps) whereas the medicinal plants were selected on the basis of their significant importance in the Ayurveda along with their frequent utilization in novel pharmaceutical preparations, preparation of traditional drug formulations, food supplements and their diverse medicinal properties. General description and economic importance of medicinal plants undertaken in the present study is given in Table 1.

Table 1: General description and economic importance of medicinal plants used in the study

Scientific name	Common name	Family	Habit	Part used	Medicinal properties	References
Acacia nilotica (L.) Willd. Ex Delile	Babul, kikar	Mimosaceae	Medium sized tree	Bark, roots, seeds, gum, pods	Used to cure diarrhea, Aphrodisiac, dressing of ulcers, Alzheimer's diseases, wound ulcers, leprosy, leucoderma, small pox, skin diseases, biliousness, burning sensation and toothache.	Ali et al. (2012)
Bacopa monnieri (L.) Pennell	Brahmi	Plantaginaceae	Perennial herb	Leaves	Improve memory capacity, intellectual activity and enhance the immune function by increasing immunoglobulin production.	Morgan and Stevens (2010) Yamada et al. (2011)

Commiphora wightii (Arn) Bhandari	Guggul	Burseraceae	Balsamiferous woody shrub	Bark, stem	Active ingredients E and Z guggulsterone in oleoresin gum have lipid and cholesterol lowering activities along with acting as anti- cancerous compound.	Kulhari et al. (2012)
Ficus religiosa L.	Peepal	Moraceae	Large deciduous tree	Bark, leaves, seeds, fruits,	Used to cure asthma, cough, sexual disorders, gonorrhea, skin diseases, scabies, hiccup, tuberculosis, fever and paralysis.	Makhija et al. (2010)
Glycyrrhiza glabra L.	Mulethi	Fabaceae	Perennial herb	Roots	Popularly used to treat ileitis, lung inflammation, peptic ulcers, leaky gut syndrome and irritable bowel syndrome.	Krausse et al. (2004)

Hemidesmus indicus L.R.Br.	Anantmul	Apocynaceae	Perennial climber	Leaves, stem	Used in inflammation, cuts, wounds, burns, skin and blood diseases, ulcers and immunological disorders.	Saravanan and Nalini (2007)
Salvadora oleoides Decne.	Jaal, Pilu	Salvadoraceae	Small tree	Leaves, roots	The leaves have anti-inflammatory, analgesic and antiulcer activities.	Yadav et al. (2008)
Terminalia bellirica (Gaertn.) Roxb.	Bahera	Combretaceae	Long tree	Leaves, bark, seed, roots, flower	Used in the treatment of gastric ulcer, constipation, general debility and piles.	Motamarri et al. (2012)
Terminalia chebula Retz.	Hartak, Harad	Combretaceae	Medium to large deciduous tree	Leaves, bark, fruits	Used to cure anemia, anorexia, leprosy, diarrhoea, bleeding piles, gout, arthritis, epilepsy abdominal pain and asthma.	Rao and Srinivas (2006)

Withania somnifera (L.) Dunal	Ashwagandha	Solanaceae	Erect branched shrub	Leaves, stem, roots, seed	Used to promote health and longevity along with revitalizing of body in debilitated conditions and increasing the capability of the individual to resist adverse environmental factors.	Weiner and Weiner (1994)

Kulhari et al.

Kulhari et al. SpringerPlus 2013 2:676, doi:10.1186/2193-1801-2-676

MATERIALS AND METHODS

Leaf and stem samples of *Acacia nilotica, Bacopa monnieri, Commiphora wightii, Ficus religiosa,Glycyrrhiza glabra, Hemidesmus indicus, Salvadora oleoides, Terminalia bellirica, Terminalia chebula* and *Withania somnifera* were collected separately from different geographical locations of Haryana and Rajasthan states of India. Sampling was done from five districts namely Jhunjhunu, Churu, Bahadurgarh, Fathebad and Mahendergarh (Table 2). The collected material was washed thoroughly with running tap water followed by washing with deionized autoclaved water to remove the dust particles and possible parasites. They were shade dried, powdered and stored in closed air tight bottles for further experimentation.

Table 2: Collection of medicinal plants from Rajasthan and Haryana states

Plant name	State	District	Latitude	Longitude	Status
Acacia nilotica	Rajasthan	Jhunjhunu	28.13° N	75.40° E	Wild
Bacopa monnieri	Haryana	Bahadurgarh	28.68° N	76.92° E	Wild
Commiphora wightii	Haryana	Mahendergarh	28.27° N	76.15° E	Wild
Ficus religiosa	Rajasthan	Jhunjhunu	28.13° N	75.40° E	Wild
Glycyrrhiza glabra	Haryana	Bahadurgarh	28.68° N	76.92° E	Wild
Hemidesmus indicus	Haryana	Fathebad	29.52° N	75.45° E	Wild
Salvadora oleoides	Rajasthan	Churu	26.60° N	75.45° E	Wild
Terminalia bellirica	Haryana	Fathebad	29.52° N	75.45° E	Wild
Terminalia chebula	Haryana	Fathebad	29.52° N	75.45° E	Wild
Withania somnifera	Haryana	Bahadurgarh	28.68° N	76.92° E	Wild

Kulhari *et al.*

Kulhari *et al. SpringerPlus* 2013 **2**:676, doi:10.1186/2193-1801-2-676

Heavy metal analysis was done according to AOAC (1995) using flame atomic absorption spectroscopy in ten medicinal plants using wet digestion method (Meena et al., 2010). Standards of Fe, Pb, Mn, Cr, Zn, Cd, Co, Hg and Ni procured from Merck, Germany, were used as reference analytes for quantitative estimation of heavy metals as well as accurate calibration and quality assurance of each analyte. The standard stock solutions (1000 ppm) were diluted to obtain working standard solutions ranging from 1 ppm to 10 ppm and stored at 4°C. An acidity of 0.1% nitric acid was maintained in all the solutions. A calibration curve was plotted between measured absorbance and concentration (ppm). All the samples were analyzed in triplicate using Flame Atomic Absorption Spectrophotometer (AA 6300 Shimadzu, Japan) with Shimadzu, Wizard software. Optimized operating parameters for FAAS (Flame Atomic Absorption Spectrophotometer) of various heavy metals are listed in Table 3.

Table 3: Operating parameters for FAAS of heavy metals

Parameters	Heavy metals								
	Mn	Cr	Pb	Fe	Cd	Ni	Co	Zn	Hg
Wavelength (nm)	279.5	357.9	283.3	248.3	228.8	232.0	240.7	213.9	253.7
Slit (nm)	0.7	0.7	0.7	0.7	0.7	0.7	0.7	0.7	0.7
Acetylene flow rate (l/min)	1.8	1.8	1.8	1.8	1.8	1.8	1.8	1.8	1.8
HCL current (mA)	6	6	6	6	6	6	6	6	6
Air flow rate (l/min)	15	15	15	15	15	15	15	15	15

Kulhari *et al.*

Kulhari *et al. SpringerPlus* 2013 **2**:676, doi:10.1186/2193-1801-2-676

For heavy metal assessment, 5 g powder of each plant sample was taken into separate crucibles for charring for 3–4 hours. After charring, the samples were placed in the Muffle furnace at 500°C for 6 hrs for ashing and were then cooled in the desiccators. Samples were digested in 20 ml mixture of concentrated acids (nitric and perchloric acid in 9:1 ratio) for 3 hours in a water bath maintained at 70°C for dissolving the contents until a clear brownish solution was obtained using wet digestion method. After cooling, these solutions were reconstituted to 20 ml volume with deionized autoclaved water. Each sample was filtered using whatmann filter paper (pore size 0.45 μ, Axiva) and stored in closed acid-washed glass vials. The stored samples were further used for analysis of heavy metals using flame atomic absorption spectroscopy. All experiments were done in triplicate for precision and accuracy of the results. Concentration of each metal was determined from absorbance value of each replicate and articulated in ppm on a dry-weight basis of the plant sample. Finally data was subjected to standard error calculation using SAS 9.3

RESULTS AND DISCUSSION

Atmosphere and soil are continuously being polluted with chemicals and heavy metals due to dynamic development of industries and motorization along with extensive use of pesticides and fertilizers. In turn, these pollutants and heavy metals are getting deposited in the plants growing in the polluted areas, which subsequently enter the human food chain via plant parts and/or extracts. Results of analysis of nine heavy metals (Mn, Cr, Pb, Fe, Cd, Co, Zn, Hg and Ni) done in stem and leaf samples of ten medicinal plant species are summarized in Tables 4 and 5 respectively. Significant difference in heavy metal concentration was observed among the plants in the two tissue samples (leaf and stem) collected from diverse geographical locations (Figures 1 and 2).

Table 4: Concentration of heavy metals in stem samples of medicinal plants

Mean heavy metal concentration in stem samples (ppm)										
S/N	Plant species	Mn	Cr	Pb	Fe	Cd	Ni	Co	Zn	Hg
1	Acacia nilotica	5.22 ± 0.1040	1.87±0.0202	0.25±0.0088	25.30±0.1464	BDL	0.45±0.0240	0.04±0.0088	2.42±0.2173	BDL
2	Bacopa monnieri	4.23±0.0744	2.62±0.0240*	2.34±0.0173	17.16±0.1763	0.02±0.0033	0.26±0.0284	0.05±0.0033	6.75±0.1223	BDL
3	Commiphora wightii	2.51±0.0176	2.87±0.0202*	0.63±0.0088	19.45±0.2334	BDL	0.55±0.0173	0.02±0.0088	7.74±0.0802	BDL
4	Ficus religiosa	4.52±0.0176	2.67±0.0371*	0.36±0.0317	24.35±0.1808	0.02±0.0010	0.64±0.0145	0.12±0.01201	4.17±0.0736	BDL
5	Glycyrrhiza glabra	4.78±0.0240	3.56±0.0305*	0.41±0.0185	21.21 ± 0.0173	0.03 ± 0.0120	0.06 ± 0.0120	0.20 ± 0.0185	5.52 ± 0.0202	BDL
6	Hemidesmus indicus	5.43 ± 0.0173	4.89 ± 0.0202*	0.66 ± 0.0352	15.42 ± 0.0176	0.03 ± 0.0145	1.07 ± 0.0057	0.23 ± 0.0881	7.13 ± 0.0115	BDL
7	Salvadora oleoides	2.85 ± 0.0218	3.68 ± 0.0202*	0.83 ± 0.0088	11.89 ± 0.0317	BDL	0.47 ± 0.0115	0.13 ± 0.0173	5.59 ± 0.0317	BDL
8	Terminalia bellirica	6.13 ± 0.0176	3.80 ± 0.0450*	0.63 ± 0.0115	16.34 ± 0.0392	0.03 ± 0.0088	0.43 ± 0.0173	0.21 ± 0.0290	5.46 ± 0.0545	BDL
9	Terminalia chebula	5.67 ± 0.0384	2.92 ± 0.0202*	0.62 ± 0.0384	16.19 ± 0.0633	0.03 ± 0.0152	0.64 ± 0.0750	0.20 ± 0.0589	6.32 ± 0.0202	BDL
10	Withania somnifera	2.39 ± 0.0218	4.93 ± 0.0185*	2.64 ± 0.0260	19.13 ± 0.0176	0.04 ± 0.0088	0.53 ± 0.0296	0.14 ± 0.0152	8.93 ± 0.0264	BDL

*Beyond permissible limit defined by WHO.

Kulhari *et al.*

Kulhari *et al. SpringerPlus* 2013 **2**:676, doi:10.1186/2193-1801-2-676

Table 5: Concentration of heavy metals in leaf samples of medicinal plants

	Mean heavy metal concentration in leaf samples (ppm)									
S/N	Plant species	Mn	Cr	Pb	Fe	Cd	Ni	Co	Zn	Hg
1	Acacia nilotica	3.16 ± 0.0371	8.73 ± 0.0260*	0.21 ± 0.0208	16.01 ± 0.0642	BDL	0.59 ± 0.0202	0.19 ± 0.0284	3.09 ± 0.0305	BDL
2	Bacopa monnieri	1.67 ± 0.0425	13.19 ± 0.0480*	0.54 ± 0.0417	14.19 ± 0.0633	0.03 ± 0.0088	0.23 ± 0.0317	0.15 ± 0.0115	4.80 ± 0.0907	BDL
3	Commiphora wightii	2.84 ± 0.0688	8.53 ± 0.0317*	0.21 ± 0.0145	13.30 ± 0.0264	BDL	0.26 ± 0.0348	0.20 ± 0.0200	2.10 ± 0.0173	BDL
4	Ficus religiosa	2.60 ± 0.0392	9.54 ± 0.0202*	0.32 ± 0.0240	12.42 ± 0.0272	0.03 ± 0.0057	0.30 ± 0.0296	0.14 ± 0.0120	3.25 ± 0.0240	BDL
5	Glycyrrhiza glabra	2.23 ± 0.0305	9.17 ± 0.0173*	0.32 ± 0.0233	11.33 ± 0.0296	0.03 ± 0.01660	0.38 ± 0.0233	0.40 ± 0.0251	3.89 ± 0.0348	BDL
6	Hemidesmus indicus	2.26 ± 0.0260	11.68 ± 0.0480*	0.25 ± 0.0088	13.64 ± 0.0115	0.16 ± 0.0260	0.32 ± 0.0173	0.16 ± 0.0251	4.18 ± 0.0212	BDL
7	Salvadora oleoides	0.89 ± 0.0458	10.48 ± 0.0440*	0.52 ± 0.0202	11.03 ± 0.0497	BDL	0.48 ± 0.0193	0.15 ± 0.0208	4.40 ± 0.0240	BDL
8	Terminalia bellirica	2.60 ± 0.0348	11.02 ± 0.0808*	0.25 ± 0.0120	13.21 ± 0.0202	0.21 ± 0.0176	0.26 ± 0.0257	0.18 ± 0.0145	3.28 ± 0.0145	BDL
9	Terminalia chebula	2.04 ± 0.0264	11.23 ± 0.0887*	0.48 ± 0.0290	14.59 ± 0.0352	0.02 ± 0.0120	0.57 ± 0.0264	0.17 ± 0.0264	2.58 ± 0.0237	BDL
10	Withania somnifera	0.34 ± 0.0152	12.34 ± 0.0458*	0.81 ± 0.0360	17.44 ± 0.0202	0.04 ± 0.0274	0.19 ± 0.0371	0.14 ± 0.0135	4.10 ± 0.0360	BDL

*Beyond permissible limit defined by WHO.

Kulhari *et al.*

Kulhari *et al.* *SpringerPlus* 2013 2:676, doi:10.1186/2193-1801-2-676

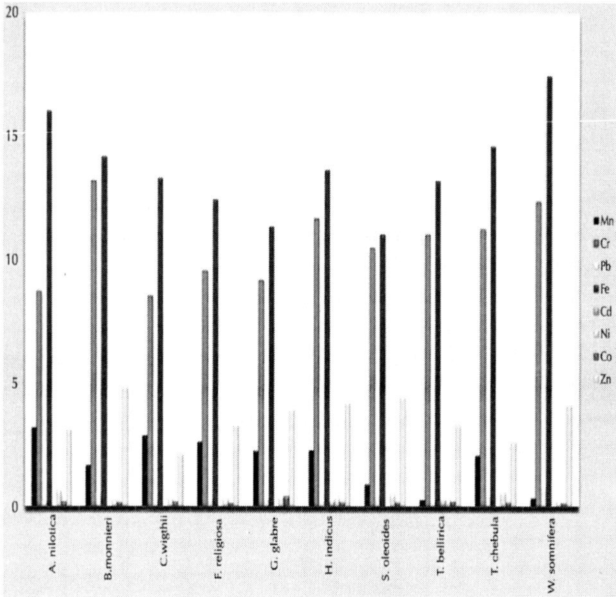

Figure 1: Heavy metal concentration in leaf samples of ten medicinal plants collected from different geographical regions.

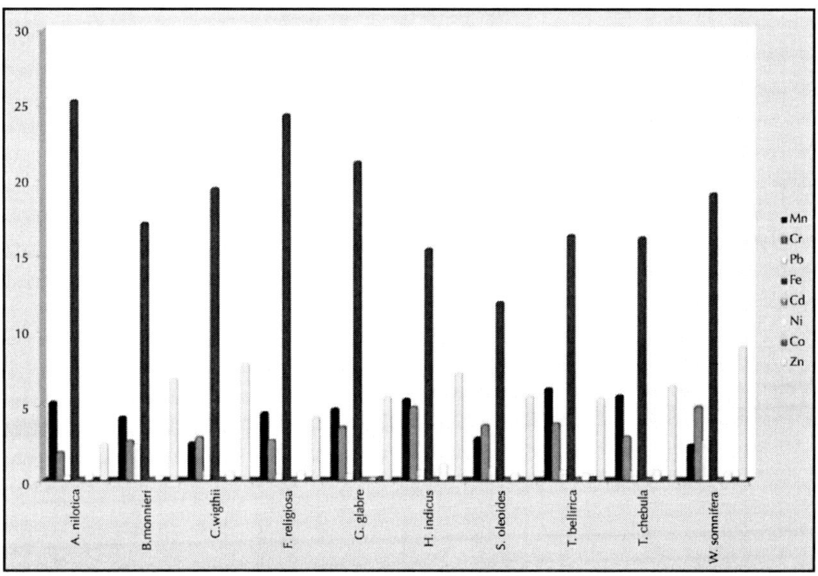

Figure 2: Heavy metal concentration in stem samples of ten medicinal plants collected from different geographical regions.

Manganese (Mn) is known as an essential trace element which acts as cofactor for many enzymes. It is less toxic than any other metal, however, can cause neurological disorders if its concentration exceeds 5 mg/m^3 due to continuous exposure to manganese dust and fumes. In the present study, a wide variation occurred in Mn concentration in stem and leaf samples, with highest content in the stem samples of *T. bellirica* (6.13 ppm) and leaf samples of *A. nilotica* (3.16 ppm) collected from Fathebad and Jhunjhunu, respectively. Mn concentration was more in the stem samples, ranging from 1.65 fold in *A. nilotica* to 7.03 fold in *W. somnifera,* compared to respective leaf samples except *C. wightii* wherein leaf samples accumulated more Mn (Figure 3). Chromium (Cr), regarded as one of the most toxic pollutants in the world is released by tanneries, steel industries, and sewage sludge applications along with alloys in motor vehicles. Its concentrations between 5–30 mg/kg are considered critical for plants as it causes heavy reduction in plant growth and yield. The permissible limit for Cr in raw herbal materials is 2.0 ppm and that for finished products is 0.02 mg/day (WHO, 2007). All the samples tested, except *A. nilotica* stem had Cr concentrations beyond the permissible limits defined by WHO. Cr concentration varied between 1.87 ± 0.0202 (A.*nilotica*) to 4.93 ± 0.0185 ppm (*W. somnifera*) in the stem samples while it varied from 8.53 ± 0.0317 (*C. wightii*) to 13.19 ± 0.0480 ppm (*B. monnieri*) in leaf samples. Cr concentration was 2.39 (*H. indicus*) to 5.03 (*B. monnieri*) fold more in leaf samples compared to respective stem samples. Lead (Pb) is highly hazardous for plants, animals and microorganisms. Continuous consumption of fertilizers, fuel combustion and sewage sludge are the major reasons leading to escalation in Pb pollution. The plant tissues tested (stem and leaf) contained extremely lower concentrations of Pb compared to the permissible limit of 10 ppm defined by WHO (2007). Maximum amount of Pb was found in stem (2.64 ± 0.0260 ppm) and leaf (0.81 ± 0.0360 ppm) samples of *W. somnifera* collected from Bahadurgarh. Pb concentration was comparable in stem and leaf samples of *A. nilotica, F. religiosa, G. glabra* and *T. chebula,* however, it was 3.0–.4.5 fold more in stem samples of *C. wightii, W. somnifera* and *B. monnieri.*

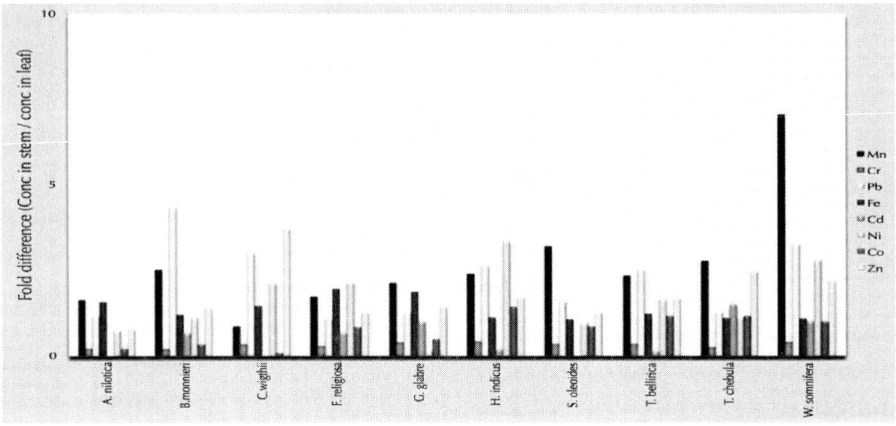

Figure 3: Fold difference in concentration of heavy metals in stem samples compared to respective leaf samples (conc. in stem sample/conc. in leaf sample).

Iron (Fe) is an important trace element and iron-protein mixtures play a vital role in metabolism in all living organisms. However, Fe overdose has been one of the leading reasons of death caused by toxicological agents in children younger than 6 years of age. Individuals demonstrate signs of gastro-intestinal toxicity after ingestion of more than 20 mg/kg iron while moderate intoxication occurs when ingestion of elemental Fe exceeds 40 mg/kg and ingestions exceeding 60 mg/kg can cause severe toxicity and may be lethal also (Spanierman, 2011). Here, highest content of Fe was found in *A. nilotica* stem samples (25.30 ± 0.1464 ppm) and *W. somnifera* leaf samples (17.44 ± 0.0202 ppm) collected from Jhunjhunu and Bahadurgarh, respectively. However, all concentrations were within the permissible limits. On the other hand, Annan et al., (2010) had reported essential element mediated toxicity due to higher accumulation of Fe (beyond the acceptable range 1000 µg/day) in *Ocimum canum, Clausena anisata* and *Rauwolfia vomitoria* collected from Ghana using wet digestion AAS. Fe concentrations were comparable (1.1–1.2 fold in *S. oleoides, W. somnifera,T. bellirica, T. chebula, H. indicus, B. monnieri* or higher (1.5–2.0 fold in *A. nilotica, C. wightii, F. religiosa, G. glabra*) in stem samples compared to respective leaf samples.

Cadmium (Cd) is another hazardous heavy metal which can cause significant reduction in plant yield at concentrations ranging from 5–30 mg/kg. Recently, it is gaining more attention due to wide occurrence in water, soil, milk, dietary products, medicinal plants and herbal products. The major sources leading to accumulation of cadmium in soil and plants are phosphate fertilizers, non-ferrous smelters, lead and zinc mines, sewage sludge application and combustion of fossil fuels. All the samples analyzed in this study had Cd concentration within the acceptable range of 0.3 ppm recommended by WHO (2007) for raw herbal material. Maximum amount of Cd was found in *W. somnifera* (0.04 ppm in both stem and leaf samples) collected from Bahadurgarh region, whereas, it remained undetected in stem samples of *A. nilotica* (Jhunjhunu), *C. wightii* (Mahendergarh) and*S. oleoides* (Churu). Cd concentration was comparable in stem and leaf samples of other plants except *T. bellirica* and *H. indicus* wherein Cd concentration was 5–7 fold more in leaf samples. Nickel (Ni) and Cobalt (Co) are trace elements required for a variety of biological processes. Ni is directly coordinated by proteins, whereas, Co is mainly used as a component of vitamin B_{12}(Zhang et al., 2009). Nickel was recognized as an allergen of the year in 2008 by the American Contact Dermatitis Society and its minimal risk level was set to 0.2 $\mu g/m^3$ for inhalation during 15–364 days, however, no limit has been set for food stuffs (Bhat et al., 2010). In the present investigation, less than 2 ppm Ni concentration was found in all the samples. Ni concentration was 2.0–3.3 folds more in stem samples in *C. wightii*, *F. religiosa*, *H. indicus* and *W. somnifera*. Concentration of Co, another major cause of contact dermatitis next only to nickel and chromium (Basketter et al., 2003), was less than 0.2 ppm in all the stem and leaf samples. Co concentration was comparable in stem and leaf samples in most of the plants studied except *A. nilotica*, *B. monnieri*, *C. wightii* and *G. glabra* where it was more in leaf samples.

Zinc (Zn) is an essential component of thousands of proteins in plants, although it is toxic in excess quantities. Maximum amount of zinc was found in stem samples of *W. somnifera* (8.93 ± 0.0264 ppm), followed by *C. wightii* (7.74 ± 0.0802 ppm) while in leaf samples highest content was found in *B. monnieri* (4.80 ± 0.0907 ppm) followed by *H. indicus* (4.18 ± 0.0212 ppm). Concentration of Zn was 2.0–3.7 folds more in stem samples of *C. wightii*, *T. chebula* and *W. somnifera*, while it was comparable in stem and leaf samples of other species tested. In Canada, 0.2 ppm has been prescribed as the limit of mercury (Hg) in raw herbal material, whereas in finished herbal products it is 0.02 mg/day. In China and Singapore however, 0.5 ppm limit has been proposed by WHO (2007). Hg concentration was below detectable limits in all the leaf and stem samples analyzed during this study. Radwan and Salama (2006) evaluated the level of Pb, Cd, Cu and Zn in various fruits and vegetables sold in Egyptian markets and found that their concentration was significantly higher but below the acceptable range in strawberries, cucumber, date and spinach. On the other hand, level of four most hazardous heavy metals Pb, Cd, Cu and Cr were found beyond the permitted limits in *Glycyrrhiza glabra, Onosma bracteatum, Viola odorata, Foeniculum vulgare,Cuminum cyminum, Coriandrum sativum* and *Zingiber officinalis* collected from southern, eastern, and western zones of Karachi, Pakistan (Hina et al., 2011).

Significant variation in metal concentration depending upon the plant species was recorded in the present study. Concentration of all heavy metals apart from Cr was found below the permissible limits in the stem and leaf samples of the ten medicinal plant species tested irrespective of their collection sites. Leaf samples had consistently more Cr compared to respective stem samples. Similarly, ten medicinal plants studied by Ajasa et al., (2004) collected from different areas of Ogbomoso revealed variable metallic levels. A higher mean level was obtained regarding Pb, Zn and Cu in *Azadirachta indica* and *Hyptis suaveolens*. Gajalakshmi et al., (2012) analyzed six heavy metals (Ni, Cu, Cr, Zn, Mn, Pb) in four medicinal plant species and revealed similar results. They reported that Cu concentration was higher in leaf samples of all the species, although lower than permissible limit, whereas Cr concentration was at higher level in all the plants, Zn and Mn were below the tolerable edge, whereas Ni and Pb were completely missing in all the analyzed samples.

The rise in Cr concentration in Bahadurgarh region (heavy industrial area located at the edge of national capital New Delhi) in the present study can be attributed to fast industrialization, higher vehicle density and rapid urbanization. Princewill-Ogbonna and Ogbonna (2011) also reported that fast urbanization in developed cities is responsible for drastic increase of metal concentration in surrounding areas and plant species growing on those sites. They reported that higher liberation of vehicular pollutants in Aba city of Nigeria exhibited variable accumulation of heavy metals in different medicinal plants as demonstrated by AAS mediated chemical profiling. Similarly, analytical profile of Pb, Cd, Cr, Ni, Sn, Zn, Mn, Cu and Fe exhibited concentrations at levels higher than permissible limits in various spice and medicinal plant species collected from exportation areas of Egypt (Abou-Arab and Abou Donia, 2000). Baranowska et al., (2002) evaluated the level of five heavy metals (Pb, Zn, Cd, Ni and Mo) in six medicinal plants through FAAS from different heavily polluted locations and concluded that all the metals were higher in concentration at Express highways, Railway stations and urban roads.

The present study concluded that bio-accumulation of heavy metals increased with rise in their trophic levels, therefore, constant utilization of the herbs and vegetables may escort its metal level in humans as also reported by Akinola et al., (2008). Highly contaminated sites are responsible for heavy metal accumulation in plant parts, utilized either raw or as finished herbal products. According to an All India Ethnobiological Survey carried out by the Ministry of Environment & Forests, Government of India, there are over 8000 species of plants being used by the people of India for medicinal purposes (Planning Commission Report, 2000). Therefore, the raw material collection sites need to be considered for their sporadic assessment for quality declaration and utilization. The medicinal plants-related trade in India estimated at Rs 5000 crores per annum with an annual growth rate of 7–15% (Joshi et al., 2004), may face serious consequences if heavy metal accumulation beyond permissible limits is detected. Therefore, assessment of these hazardous metals in raw herbal material and finished products must be undertaken compulsorily. Recently, studies have been initiated for detection of heavy metals in many medicinal plants (Hussain et al., 2011; Baye and Hymete, 2010; Meena et al., 2010) utilizing AAS. Concentration of Pb and Cd was more than the permissible limits recommended by WHO for some medicinal plants analyzed by Baye and Hymete (2010). Instrumental neutron activation analysis and AAS of five heavy metals (Cr, Ni, Cd, As and Pb) and seven essential metals (V, Mn, Fe, Co, Cu, Zn, Se) from Khetri copper mine and those in fertile soils of Haridwar, India exhibited a wide variation in heavy metal concentration sketch. Concentrations of Cu, Cr, Cd and Pb were higher in Khetri copper mine as compared to Haridwar samples, whereas Ni, As and other essential metals showed little variation in both the collection sites (Maharia et al., 2010). It was also reported that *Withania somnifera* showed highest bio-accumulation of metals as compared to *Ocimum sanctum*, *Cassia fistula*, and *Azadirachta Indica*.

CONCLUSIONS

Medicinal plants are sources of a large number of active principles of herbal and modern medicine. Indian people have a tremendous passion for medicinal plants and use them for a wide range of health related applications from a common cold to cancer and treatment of poisonous snake bites to a cure for genetic disorders like muscular dystrophy. India has one of the richest herbal medical cultures in the world that is of tremendous contemporary relevance ensuring health security to millions of people. However, continuous increase in environmental pollution is leading to build up of these pollutants including heavy metals in the plant parts which eventually enter the human food chain. Therefore, regular screening of raw material is must to check the levels of these pollutants in the plant parts and extracts before using them for human consumption. The present investigation clearly demonstrated the variation in heavy metal concentration depending upon the collection sites. Different countries have proposed different national limits for various heavy metals in raw herbal materials e.g the minimum permissible limits for arsenic, lead, cadmium, mercury and chromium are 5.0, 10.0, 0.3, 0.2 and 2.0 ppm respectively in Canada while the same are 2.0, 10.0, 1.0, 0.5 ppm respectively in China, and the limit has not yet been prescribed for chromium (WHO, 2007). However, the permissible limits have not been prescribed for many metals which are considered micronutrients. The developed AAS technique is a precise, specific and accurate method for estimation of heavy metal content in stem and leaf samples of these medicinal plants. Assessment of heavy metals in medicinal plants will pave the way for excluding extensively polluted environmental sites for collection of raw materials required for herbal drug preparation.

AUTHORS' CONTRIBUTION

AK, AS and SB conducted the experiments, SS analyzed the data, AC and RKK designed the research experiments and arranged the research facilities. All authors read and approved the final manuscript.

ACKNOWLEDGEMENT

AK thankfully acknowledges the financial assistance provided by Department of Biotechnology, Government of India, New Delhi, under the project sanctioned vides order no. BT/PR10526/NDB/51/164/2008.

REFERENCES

1. Abou-Arab AA, Abou Donia MA (2000) Heavy metals in Egyptian species and medicinal plants and the effect of processing on their levels. J Agric Food Chem 48:2300-2304
2. Ajasa MA, Bello OM, Ibrahim OM, Ogunwande AI, Olawore ON (2004) Heavy trace metals and macronutrients status in herbal plants of Nigeria. Food Chem 85:67-71
3. Akinola MO, Njoku KL, Ekeifo BE (2008) Determination of Lead, Cadmium and Chromium in the tissue of an economically important plant grown around a textile industry at Ibeshe, Ikorodu Area of Lagos State, Nigeria. Adv Environ Biol 2:25-30
4. Ali A, Akhtar N, Khan BA, Khan MS, Rasul A, Zaman SUZ, Khalid N, Waseem K, Mahmood T, Ali L (2012) *Acacia nilotica*: a plant of multipurpose medicinal uses. J Med Plant Res 6:1492-1496
5. Annan K, Kojo AI, Cindy A, Asare-Nkansah S, Tunkumgnen BM (2010) Profile of heavy metals in some medicinal plants from Ghana commonly used as components of herbal formulations. Pharmacognosy Res 2:41-44
6. AOAC (1995) Official methods of analysis of AOAC International. AOAC International, Arlington, USA.
7. Baranowska I, Srogi K, Włochowicz A, Szczepanik K (2002) Determination of heavy metal contents in samples of medicinal herbs. Pol J Environ Stud 11:467-471
8. Basketter DA, Angelini G, Ingber A, Kern PS, Menné T (2003) Nickel, chromium and cobalt in consumer products: revisiting safe levels in the new millennium. Contact Dermatitis 49:1-7
9. Baye H, Hymete A (2010) Lead and Cadmium accumulation in medicinal plants collected from environmentally different sites. Bull Environ Contam Toxicol 84:197-201

10. Bhat R, Kiran K, Arun AB, Karim AA (2010) Determination of mineral composition and heavy metal content of some nutraceutically valued plant products. Food Anal Methods 3:181-187
11. Gajalakshmi S, Iswarya V, Ashwini R, Divya G, Mythili S, Sathiavelu A (2012) Evaluation of heavy metals in medicinal plants growing in Vellore District. Eur J Exp Biol 5:1457-1461
12. Gupta S, Porwal MC, Roy PS (2004) Indigenous knowledge on some medicinal plants among the Nicobari Tribe of Car Nicobar Island. Indian J Tradit Know 3:287-293
13. Hina B, Rizwani GH, Naseem S (2011) Determination of toxic metals in some herbal drugs through atomic absorption spectroscopy. Pak J Pharm Sci 24:353-358
14. Hussain I, Ullah R, Khurram M, Ullah N, Baseer A, Khan FA, Khan N, Khattak MR, Zahoor M, Khan J (2011) Heavy metals and inorganic constituents in medicinal plants of selected districts of Khyber Pakhtoonkhwa, Pakistan. Afr J Biotechnol 42:8517-8522
15. Joshi K, Chavan P, Warude D, Patwardhan B (2004) Molecular markers in herbal drug technology. Curr Sci 87:159-165
16. Kloke A, Sauerback DR, Vetter H (1984) The contamination of plants and soils with heavy metals and the transport of metals in terrestrial food chains. In: Nriagu JO (ed) Changing Metal Cycles and Human Health, Springer-Verlag, Berlin. pp 113-141
17. Krausse R, Bielenberg JBW, Ullmann U (2004) *In vitro* anti-*Helicobacter pylori* activity of extractum liquiritiae, glycyrrhizin and its metabolites. J Antimicrob Chemother 54:243-246
18. Kulhari A, Sheorayan A, Kalia S, Chaudhury A, Kalia RK (2012) Problems, progress and future prospects of improvement of *Commiphora wightii* (Arn.) Bhandari, an endangered herbal magic, through modern biotechnological tools: a review. Genet Resour Crop Evol 59:1223-1254
19. Kunle OF, Egharevba HO, Ahmadu PO (2012) Standardization of herbal medicines-a review. Int J Biodivers Conserv 4:101-112
20. Lee SM, Lee WY (2002) Determination of heavy metal ions using conductometric biosensor based on sol-gel-immobilized urease. Bull Kor Chem Soc 23:1169

21. Maharia RS, Dutta RK, Acharya R, Reddy AVR (2010) Heavy metal bioaccumulation in selected medicinal plants collected from Khetri copper mines and comparison with those collected from fertile soil in Haridwar, India. J Environ Sci Health B 45:174-181
22. Makhija IK, Sharma IP, Khamar D (2010) Phytochemistry and pharmacological properties of *Ficus religiosa*: an overview. Ann Biol Res 1:171-180
23. Meena AK, Bansal P, Kumar S, Rao MM, Garg VK (2010) Estimation of heavy metals in commonly used medicinal plants: a market basket survey. Environ Monit Assess 170:657-660
24. Morgan A, Stevens J (2010) Does *Bacopa monnieri* improve memory performance in older persons? Results of a randomized, placebo-controlled, double-blind trial. J Altern Complement Med 16:753-759
25. Motamarri SN, Karthikeyan M, Kannan M, Rajasekar S (2012) *Terminalia bellerica* Roxb–a phytopharmacological review. Int J Res Pharma 3:96-99
26. Nies DH (1999) Microbial heavy metal resistance. Appl Microbiol Biotechnol 51:730-750
27. Nwoko CO, Mgbeahuruike L (2011) Heavy metal contamination of ready-to-use herbal remedies in south eastern Nigeria. Pakistan J Nutr 10:959-964
28. Olajire AA, Ayodele ET (2003) Study of atmospheric pollution levels by trace elements analysis of tree bark and leaves. Bull Chem Soc Ethiop 17:11-17
29. Planning Commission Report (2000) Report of the Task Force on conservation and sustainable use of medicinal plants. Planning Commission, Government of India. 1-194
30. Princewill-Ogbonna IL, Ogbonna PC (2011) Heavy metal content in soil and medicinal plants in high traffic urban area. Pakistan J Nut 10:618-624
31. Radwan AM, Salama AK (2006) Market basket survey for some heavy metals in Egyptian fruits and vegetables. Food Chem Toxicol 44:1273-1278
32. Rao NK, Srinivas N (2006) Antidiabetic and renoprotective effects of the chloroform extract of*Terminalia chebula* Retz. seeds

in streptozotocin-induced diabetic rats. BMC Compl Alternative Med 6:17

33. Rao MR, Palada MC, Becker BN (2004) Medicinal and aromatic plants in agro-forestry systems. Agroforestry Syst 61:107-122
34. Saravanan N, Nalini N (2007) Impact of *Hemidesmus indicus* R.Br. extract on ethanol-mediated oxidative damage in rat kidney, Redox report. Free Radic Res Comm 12:229-235
35. Spanierman CS (2011) Iron toxicity in emergency medicine.
36. http://emedicine.medscape.com/article/815213-overview#showall
37. Szeto M (2007) *Coriolus versicolor* extracts: relevance in cancer management. Curr Oncol 14:41-47
38. Weber G, Konieczy ski P (2003) Speciation of Mg, Mn and Zn in extracts of medicinal plants. Anal Biochem 375:1067-1073
39. Weiner MA, Weiner J (1994) *Ashwagandha* (Indian ginseng). In: Herbs that Heal. Quantum Books, Mill Valley, CA. pp 70-72
40. WHO (2007) Guidelines for assessing quality of herbal medicines with reference to contaminants and residues. World Health Organization, Geneva.
41. Yadav JP, Saini S, Kalia AN, Dangi AS (2008) Hypoglycemic and hypolipidemic activity of ethanolic extract of *Salvadora oleoides* in normal and alloxan-induced diabetic rats. Indian J Pharmacol 40:23-27
42. Yamada K, Hung P, Park TK, Park PJ, Limb BO (2011) A comparison of the immunostimulatory effects of the medicinal herbs Echinacea, Ashwagandha and Brahmi. J Ethnopharmacol 137:231-235
43. Zhang Y, Rodionov DA, Gelfand MS, Gladyshev VN (2009) Comparative genomic analyses of nickel, cobalt and vitamin B_{12} utilization. BMC Genomics 78:1-26

Chapter 7

Immobilization of Leachate's Heavy Metals Using Soil-Zeolite Column

Sayyed Mohammad Javad Mirzaei[1,2], Manouchehr Heidarpour[1], Sayyed Hassan Tabatabaei[2], Payam Najafi[3], and Seyyed Ebrahim Hashemi[4]

[1]Department of Irrigation, Faculty of Agriculture, Isfahan University of Technology, Isfahan, Iran

[2]Department of Water Engineering, Faculty of Agriculture, Shahrekord University, Shahrekord, Iran

[3]Department of Soil Science, Faculty of Agriculture, Khorasgan Branch, Islamic Azad University, Isfahan, Iran

[4]Department of Water Engineering, College of Agriculture, Isfahan University of Technology, Isfahan, Iran

ABSTRACT

Introduction

Soil hazardous heavy metal pollution is increasingly a decisive problem all over the world. One of the problems associated with organic fertilizer factories is the discharge of leachate containing solid wastes. The leachate percolates through the soil and where there is no proper leachate filtration, this discharge could be potentially a primary pollution source of soil and water resources. Iran has limited water resources and oppositely large number of unused lands. Therefore, properly managed land treatment may be a suitable solution for the filtration of leachate before it enters the environment or is reused as irrigation water. To investigate the potentiality of soil-zeolite in the immobilization of heavy metals and prevention of groundwater contamination, a lysimeter study was performed with two types of soil texture (clay loamy and loamy sand) and two levels of zeolite and a blank (0, 5, and 10%).

Results

The results showed that soil with a clay loamy texture can adsorb heavy metals such as Cr^{3+}, Pb^{2+}, Ni^{2+}, and Cd^{2+} from influent leachate stronger than soil with loamy sand texture. Application of zeolite had a significant effect ($p < 0.05$) on Pb^{2+}, Ni^{2+}, and Cr^{3+} concentration in the effluent. Increasing applied zeolite reduced the concentrations of these heavy metals in the effluent. Furthermore, application of zeolite and irrigation with leachate decreased the bulk density, hydraulic conductivity, and infiltration of the soil in all treatments.

Conclusion

Briefly, this research indicated that land treatment is a good way to reduce leachate electrical conductivity (EC) and elimination of leachate heavy metals.

INTRODUCTION

The moisture content of urban solid wastes in Iran is very high. Therefore, a large volume of leachate is produced during the process of converting wastes into compost. The leachate contains large amounts of organic matter, plant nutrients, and soluble salts as well as some heavy metals Kalbassi and Gandomkar (1997). Hence, it is necessary to refine the leachate of urban solid wastes before it has the chance to enter the environment or is reused.

Conventional wastewater treatment technologies are cost-intensive and often pose financial constraints for developing countries. Therefore, considerable attentions have been directed toward the design and development of low-cost wastewater management technologies with recycling and reuse benefits (Ackley and Yang 1991; Thawale et al. 2006).

Land treatment is accomplished by biological, chemical and physical interactions in the soil matrix at the surface layers, which comprises the most active zones of the soil. Land treatment is a practice of spreading contaminated soils or wastes on the surface of the ground to facilitate natural microbial degradation of contaminants. This process is defined as the application of municipal and industrial wastewater in the land at a controlled rate. The purpose of this treatment system is to obtain beneficial use of these materials, to improve environmental quality, and to achieve treatment goals in a cost-effective and environmentally sound manner. The direct discharge of effluent in soils can be used as a primary or secondary treatment and offers opportunities for groundwater recharge and wastewater storage. The effects of wastewater on soil, subsurface water, and plants completely depend on the type and content of the wastewater, the soil texture, and the prevailing conditions in the soil Bhardwaj et al. (2007).

In recent years, natural substances such as zeolite have been used to adsorb heavy metals from water Ouki and Kavannagh (1999) and from soil to decrease their availability to plants (Mondale et al. 1995; Inglezakis et al. 2002). Zeolite is a natural material that has been studied extensively for its remediation potentiality of hazardous heavy metal-polluted soils Shanableh and Kharabsheh (1996). Compared to other techniques, zeolite application is fast, clean, and inexpensive Mahabadi et al. (2007). Zeolite consists of crystallized silicate with a porous structure. This material has been the subject of great interest in a wide array of applications Breck (1974). Zeolites, as natural cation exchangers, are suitable to remove toxic cations Mondale et al. (1995; Inglezakis et al. 2002). Among the natural zeolites, clinoptilolite seems to be the most efficient ion exchanger and ion-selective material Bittell and Miller (1974) for removing and stabilizing heavy metals Echeverria et al. (1998). It has a high specific surface area and superior capability for adsorption, cation exchange, and catalyzation Leppert (1990). Application of zeolite in the soil can retard the migration of leachate and, as a result, allow the exchange and/or decomposition of heavy metals Shi et al. (2009). Leachate land treatment can affect the physical or chemical properties of the soil, such as hydraulic conductivity, infiltration, electrical conductivity (EC), the presence of heavy metals, and the cation and anion content.

Iskander et al. (2011) showed the possibility of using natural zeolite and bentonite as slow release fertilizers for Zn and Mn and also as soil pollution preventers. Zn in the case of zeolite showed the lowest desorbed percentage particularly at high levels of sorbed Zn. After three successive extractions, only 74.7% was readily extractable by diethylenetriaminepentaacetic acid (DTPA) excluding 25.3% Zn retained by the mineral whereas in the bentonite case, 82.26% of sorbed Zn was readily extractable by DTPA excluding 17.74% Zn retained by the mineral. In the case of Mn, 84.63% of Mn sorbed by zeolite was readily extractable by DTPA apart from 15.37% Mn retained by the mineral. In the case of bentonite, 89.79% of adsorbed Mn was readily extractable by DTPA, exclusive of 10.21% Mn retained by the mineral. In all treatments three successive extractions were implemented. Tabatabaei et al. (2012) demonstrated that adding zeolite to clay loam had no significant effect on the chemical oxygen demand. Sandy loam with zeolite had high significant impact on the removal capacity of total coliform (TC); oppositely, clay loam had no impact on the removal capacity of TC. Clay loam implemented better performance than the sandy loam soil on Na, Ca, and Mg adsorption.

The focus of this paper is to investigate the potential of soil-zeolite in the immobilization of heavy metals and the effect of leachate on physical properties of soil in the presence of zeolite.

METHODS

Zeolite Sampling

The natural zeolite (Clinoptilolite zeolite) used in this study originated from Semnan province, Iran. The cation exchange capacity (CEC) value of this sample was 100 meq $100 \cdot g^{-1}$. Table 1 shows the chemical and physical characteristics of the zeolite sample used in this experiment.

Table 1: Chemical and physical characteristics of the zeolite sample

Parameter	Value/description
P2O5 (%)	0.01
MnO (%)	0.04
Fe2O3 (%)	1.30
TiO2 (%)	0.21
K2O (%)	3.12
Na2O (%)	2.01
MgO (%)	0.80
CaO (%)	3.10
Al2O3 (%)	11.80
SiO2 (%)	66.50
LOIa	12.05
Color	Light green
Mineral (%)	85 to 95
CEC (mEq 100·g–1)	100
ρsb (g·cm–3)	2.40
ρbc (g·cm–3)	1

[a]Loss of ignition, [b]Particle density, [c]Bulk density.

Mirzaei *et al.*

Mirzaei et al. *International Journal Of Recycling of Organic Waste in Agriculture* 2013 2:20, doi: 10.1186/2251-7715-2-20

Site Description

The study was conducted at Organic Fertilizer factory located in the east of Isfahan at the central part of Iran. In the study area, the mean annual temperature, rainfall, and humidity are 17°C, 134 mm, and 38%, respectively. A complete randomized block design experiment with six treatments was applied (A0: loamy sand soil, A5: loamy sand soil mixed with 5% zeolite, A10: loamy sand soil mixed with 10% zeolite, B: clay loamy soil, B5: clay loamy soil mixed with 5% zeolite, and B10: clay loamy soil mixed with 10% zeolite) and three replications were performed in 18 PVC soil columns filled with treatment soils (60-cm diameter and 100-cm height with a drain pipe at the bottom). The physical characteristics of the two types of soil used in this study are listed in Table 2. A 10-cm layer of fine sand was placed in the bottom of each lysimeter as a filter, and the next 60 cm was filled with the treatment.

Table 2: Physical characteristics of the soil applied to the lysimeters

Characteristic	Loamy sand	Clay loam
Hydraulic conductivity (m·day−1)	3.02	1.58
Electrical conductivity (dS·m−1)	0.34	0.38
pH	6.85	6.54
Organic carbon (%)	0.10	0.48

Mirzaei *et al.*

Mirzaei *et al. International Journal of Recycling of Organic Waste in Agriculture* 2013 **2**:20, doi: 10.1186/2251-7715-2-20

Leachate from Isfahan Organic Fertilizer factory was used as the influent in the lysimeters. The physicochemical properties of the leachate are given in Table 3. Influent application was scheduled bi-weekly, and the total number of influent application through the study period was 16. The amount of applied influent for all treatments was 5 cm in each influent application event. The total amount of applied leachate for each lysimeter was approximately 226 L. Figure 1 shows a schematic drawing of the lysimeter.

Table 3: Physicochemical analyses of influent leachate

SAR	pH	EC	Cl^-	HCO_3^-	Na^+	Ca^{2+}	Mg^{2+}	TDS	TSS	Cr^{3+}	Cd^{2+}	Pb^{2+}	Ni^{2+}
		$dS \cdot m^{-1}$	$Mg \cdot L^{-1}$										
11.88	4.70	39.32	10,472	32,395	5,175	8,320	3,660	58,400	9,026	1.07	0.12	1.25	2.62

Mirzaei *et al.*

Mirzaei *et al. International Journal of Recycling of Organic Waste in Agriculture* 2013 **2**:20, doi: 10.1186/2251-7715-2-20

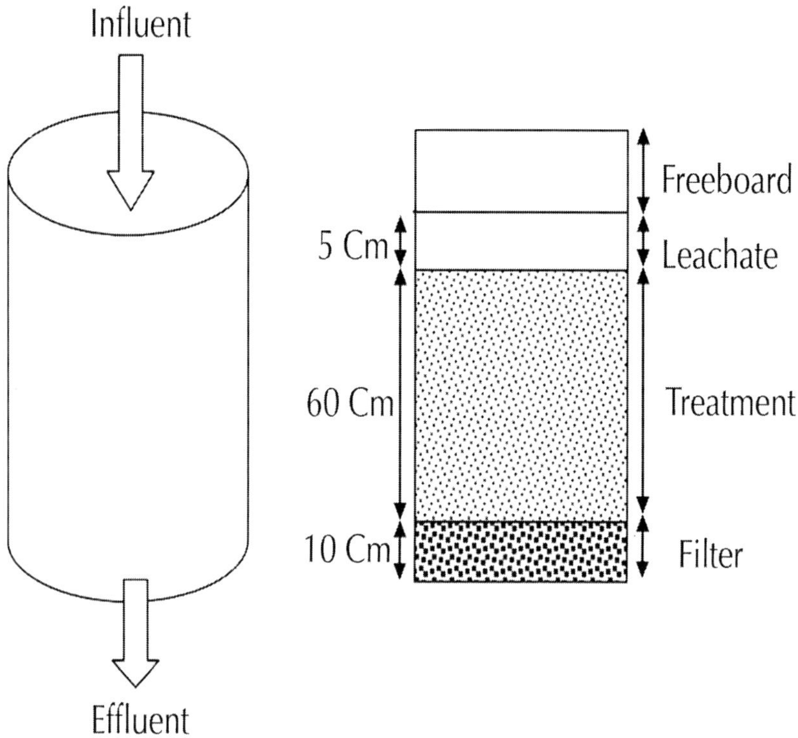

Figure 1: A schematic drawing of the lysimeter.

Data Collection

The leached solution was collected weekly from each lysimeter by opening the drainage valve located at the end of drainage pipe. The drainage samples were subjected to analysis for electrical conductivity (EC) and pH value. EC was determined using a conductivity meter, and pH was measured with a pH meter. The concentrations of cadmium (Cd^{2+}), nickel (Ni^{2+}), chromium (Cr^{3+}), and lead (Pb^{2+}) in effluent samples were specified by standard methods during the study period APHA (1995). Heavy metal concentrations were measured using an atomic adsorption spectrometer (PerkinElmer 3030, Waltham, MA, USA). At the end of the study, undisturbed soil samples were collected from each lysimeter using a core sampler. The bulk density and particle density of the soil samples were calculated using the cylinder and picnometer methods, respectively Klute (1986). The hydraulic conductivity of the soil was identified in each lysimeter using constant (for clay loamy soil) and falling head (for loamy sand soil) methods Shanableh and Kharabsheh (1996).

Data Analysis

The percentage of removed heavy metal was calculated by the following formula:

$$E=100*(C_i-C_e)/C_i, \quad (1)$$

where C_i is the metal concentration in the influent leachate sample in $mg \cdot L^{-1}$ and C_e is the metal concentration in the effluent leachate sample in $mg \cdot L^{-1}$.

All analyses were performed using SAS statistical analysis SAS (1987). Separation of means was performed using LSD at $p < 0.05$.

RESULTS AND DISCUSSION

EC

EC is typically used to indicate salt concentration in solution. Figure 2 shows the value of the EC for the influent (leachate from the fertilizer factory) and the effluent of each treatment during the study. The EC of the influent was measured within the range of 30 to 50 $dS \cdot m^{-1}$ during the experiment, which indicates the complex ion composition of this kind of wastewater. Application of zeolite had no effect on the effluent EC in the loamy sand lysimeter. In the first 4 weeks, the effluent EC calculated in these lysimeters was lower than the influent EC (Figure 1). During the mentioned weeks, the maximum difference between the influent and effluent EC values was 17 $dS \cdot m^{-1}$. After this period, the effluent EC values measured in these lysimeters increased, and after week 9, it was higher than the influent EC value. This result could possibly be due to accumulation of salts in the soil during the first weeks and leaching these elements at the end of study. Hatt et al. (2007) demonstrated the fact that concentrations of heavy metals and phosphorus in outflow may increase along with time, because of the slow washing of fine particles through the filter and/or desorption of the elements due to changing pH and oxygen levels as the filter clogs. During the study period, the EC of the effluent from the clay loam lysimeters was lower than the EC of the influent, which was between 3.5 to 27 $dS \cdot m^{-1}$.

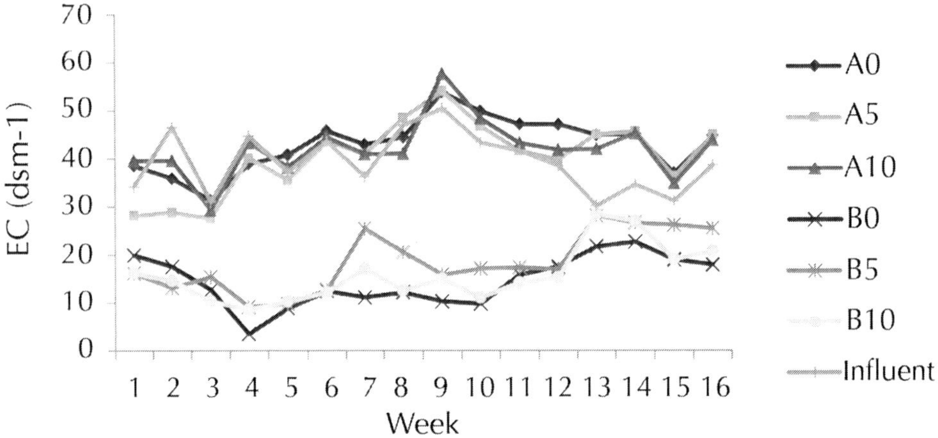

Figure 2: Changes in the influent and effluent EC in all treatments.

The results also showed that soil type had a significant effect ($p < 0.05$) on the effluent EC value and the lowest effluent EC value was found in lysimeters with clay loamy soil texture. The mean value of the effluent EC for clay loamy and loamy sand soil texture were 16.28 and 41.87 $dS \cdot m^{-1}$, respectively. Because clay loamy soil has a higher cation exchange capacity, it was able to adsorb positive cations such as sodium and, as result, decrease the value of the effluent EC up to more than 60%. The level of zeolite added to the soil had a significant effect ($p < 0.05$) on the effluent EC value. In more detail, by adding 5% and 10% natural zeolite to the soil, the mean effluent EC values were 30.3 and 28.13 $dS \cdot m^{-1}$, respectively. To clarify this phenomenon, it should be said that natural zeolite has a very open framework with a network of pores that offers a large surface area for trapping and exchanging valuable nutrients. Natural zeolites adsorb salinity ions, which enter cavities of the material and consequently, the salinity levels of the effluent are decreased Turan (2008).

pH

Figure 3 demonstrates fluctuations in effluent pH values during the study for all treatments. The mean value of influent pH during the study period was equal to 4.73. The mean effluent pH values for loamy sand and clay loamy soil were 5.48 and 7.77, respectively. There was a significant difference between the mean effluent pH values of loamy sand and clay loamy soils. Soil pH affects all chemical, physical, and biological properties of the soil Brady and Weil (2008). The effect of soil pH on specific microorganisms, soil microbial biomass, and microbial activity and, more recently, on microbial community structure have been already investigated Pietri and Brookes (2008). In previous studies, it was shown that clay loamy soil decreases the influent acidity and effectively increases the effluent pH. They also discovered that a significant effect ($p < 0.05$) on effluent pH values is measured based on the volume of applied zeolite. Along with zeolite application increase, the pH value of the effluent also increased. The mean of effluent pH values for 5% and 10% zeolite were equal to 6.49 and 6.75, respectively. Applying 10% zeolite resulted in the increase in the effluent pH values of loamy sand and clay loamy soil from 4.73 to 5.31 and 7.92, respectively. Zeolite has a high cation exchange capacity and is able to adsorb positive cations such as hydrogen and sodium. Therefore, with increasing zeolite content of the soil, the effluent pH increased as well. This finding is consistent with previous studies (Shanableh and Kharabsheh 1996; Mahabadi et al. 2007; Noori et al. 2007; Perez-Caballero et al. 2008; Ahmed et al. 2010).

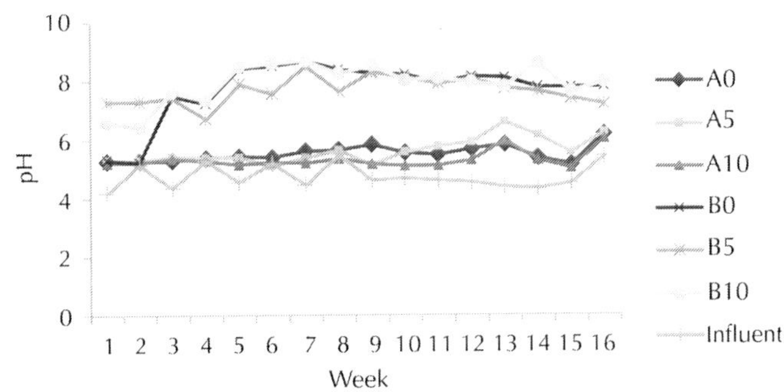

Figure 3: Changes in the pH values of the effluent during the study for all treatments.

Heavy Metal Reduction

The stabilization of Cd^{2+}, Ni^{2+}, Pb^{2+}, and Cr^{3+} in control and treatment lysimeters was measured. Table 4 shows the concentrations of Cd^{2+}, Ni^{2+}, Pb^{2+}, and Cr^{3+} in the effluents obtained from different treatments. As shown in this table, soil texture had a significant effect on the reduction of Cd^{2+}, Ni^{2+}, Pb^{2+}, and Cr^{3+} in the effluent. Furthermore, lower concentration of these heavy metals in the effluent obtained from clay loamy lysimeters was seen comparing loamy sand lysimeters. This result is likely due to the particle size, specific surface area and CEC of the soils. Clay loam has a smaller particle size and a higher specific surface area than loamy sand soil; and because of that, it was able to more effectively remove these elements from the influent than the loamy sand soil.

Table 4: Mean concentration of heavy metal in the treatment effluents

Soil texture	Zeolite application (%)	Cd	Ni	Cr	Pb
		mg·L−1			
Loamy sand	0	0.11a*	3.09a	0.82a	1.13a
Loamy sand	5	0.11ab	2.87b	0.81a	1.12ab
Loamy sand	10	0.10b	2.69c	0.64b	1.06b
Clay loam	0	0.03c	0.47d	0.06c	0.31d
Clay loam	5	0.03c	0.53d	0.03c	0.38c
Clay loam	10	0.03c	0.50d	0.03c	0.37c

*Values in any column followed by the same letter do not differ at the 0.05 significance level.

Mirzaei *et al.*

Mirzaei *et al. International Journal of Recycling of Organic Waste in Agriculture* 2013 2:20, doi: 10.1186/2251-7715-2-20

The lysimeter effluents' changes in Cd^{2+} concentration in all treatments is revealed in Figure 4. According to Table 4, soil texture had a significant effect on Cd^{2+} concentration. The mean percentages of Cd^{2+} reduction for clay loam and loamy sand soils were equal to 76.41% and 12.24%, respectively. Additionally, application of 5% and 10% zeolite had no significant effect on Cd^{2+} adsorption (Table 4). Three weeks after starting the experiment, results uncovered that the loamy sand soil lysimeters with 5% and 10% zeolite could adsorb 28.6% and 52.4% of Cd^{2+} from the influent, respectively. Our results also indicate that during the first weeks of the study, application of zeolite could decrease the leaching of Cd^{2+}. These results could be related to the low influent pH values, suggesting that pH value is an important factor which effects Cd^{2+} leachability Shanableh and Kharabsheh (1996).

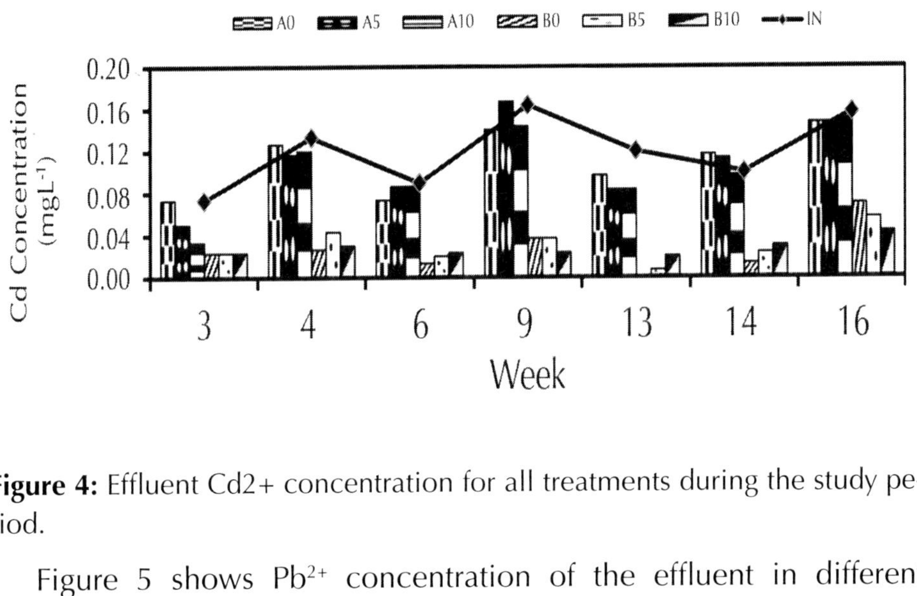

Figure 4: Effluent Cd2+ concentration for all treatments during the study period.

Figure 5 shows Pb^{2+} concentration of the effluent in different treatments. Except the loamy sand lysimeter without zeolite application (A_0), the Pb^{2+} concentration of effluent during the first 3 weeks in all treatments was similar. In all treatments except A_0 (30%), removed Pb^{2+} percentage from the influent was approximately 50%. While application of zeolite significantly ($p < 0.05$) increased the Pb^{2+} sorption in the loamy sand lysimeters, no significant difference was found ($p < 0.05$) between application of 5% and 10% zeolite in clay loamy soil in terms of Pb^{2+} concentrations in the effluent (Table 4). The mean percentages of Pb^{2+} removal identified for A_0, A_5, A_{10}, B_0, B_5, and B_{10} were 10.4%, 14.9%, 17.2%, 75%, 68%, and 69% respectively.

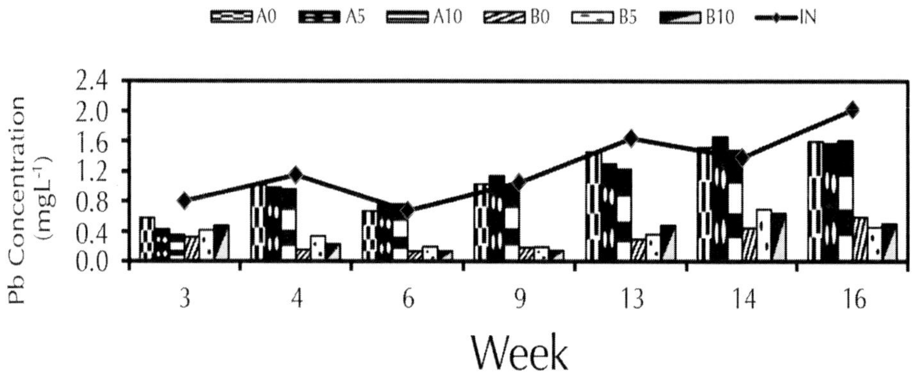

Figure 5: The Pb2+ concentration of the effluent from different treatments.

Application of zeolite had a significant effect ($p < 0.05$) on effluent Ni^{2+} and Cr^{3+} concentrations in loamy sand soil (Table 4). The results showed that application of 10% zeolite reasonably reduced Cr^{3+} and Ni^{2+} concentrations. The maximum concentrations of Ni^{2+} and Cr^{3+} in the effluent were observed in lysimeters without application of zeolite. The mean concentrations of Ni^{2+} in effluents obtained from lysimeters with 0%, 5%, and 10% zeolite were 1.78, 1.7, and 1.6 $mg \cdot L^{-1}$, respectively. The mean concentrations of Cr^{3+} in effluents obtained from lysimeters with 0%, 5%, and 10% zeolite were 0.437, 0.423, and 0.334 $mg \cdot L^{-1}$, respectively (data are not shown).

Leachate from an organic fertilizer factory was used as the influent that consisted of many different cations. Therefore, the results of this study were different from previous studies which used solutions with specific elements.

There are some experimental reports from the fact that increasing zeolite content results in lower Cd^{2+} and Pb^{2+} concentrations in the leaching solution Lin et al. (1998; Oste et al. 2002; Mahabadi et al. 2007); however, in this study, the application of zeolite only increased the adsorption during the first 3 weeks. This inconsistent result could be attributed to the chemical characteristics of organic fertilizer factory leachate that was used as the influent in this research. The pH value of the influent used in this study was very low compared to the previous studies Mahabadi et al. (2007). It has been confirmed that sorption is mainly dominated by ion exchange, as H^+ ions compete Pb^{2+} ions for ion exchange sites (Sheng et al. 2009; Wang et al. 2009; Yang et al.

2011). Thus, the higher the pH value (i.e., lower H^+ concentration), the higher the sorption of Pb^{2+}, Cr^{3+}, and Ni^{2+} on the surface of zeolite. Yang et al. (2011) demonstrated that the sorption of Pb^{2+} in low pH can be attributed to ion exchange between Pb^{2+} and H^+/Na^+ on the surface of NKF-6 zeolite. Nevertheless, the presence of other cations such as K^+ and Na^+ in the solution influences the sorption of heavy metals. The competitive cations hinder the exchange of heavy metals through zeolite, which is a well-known phenomenon in ion exchange systems Yang et al. (2011).

With the exception of the first three influent applications and in contrast with clay loamy soil treatments (B_0, B_5, and B_{10}), the loamy sand soil treatments could not adsorb Pb^{2+}. The CEC of the loamy sand soil and zeolite were saturated with cations and anions such as Na^+, Ca^{2+}, Cl^-, which existed in the influent upon first leachate applications. Exchangeable ions, such as Na^+, K^+, Ca^{2+}, and Mg^{2+} commonly occupy clinoptilolite Ackley and Yang (1991). These cations are exchangeable with certain cations in the solution, such as Pb^{2+}, Cd^{2+}, Zn^{2+}, and Mg^{2+} Erdem et al. (2004). Some investigators have used a natural zeolite additive to reduce Pb^{2+}, Cd^{2+}, and Ni^{2+} leaching from a soil contaminated with the mixtures of the three metals. The results from the repeated leaching column experiments confirmed the selectivity of the additive, which brought about a satisfactory leaching reduction of Cd^{2+} and Pb^{2+} (Ackley and Yang 1991; Shi et al. 2009; Zhang et al. 2010). Wang et al. (2009) exhibited that K^+, Na^+, and Li^+ (as competitive ions) decreased the Pb^{2+} sorption at $pH < 6$. The influence of these cations on the sorption of heavy metals can also be attributed to the hydrated radius. K^+ is smaller than those of the other two cations and therefore, the effect of K^+ on reduction of Pb (II) adsorption is more obvious than those of Na^+ and Li^+. Li et al. (2009) investigated the effect of Li^+, Na^+, and K^+ on Cu sorption in GMZ bentonite, and similar results were found.

Bulk Density

Table 5 shows the comparison between the bulk densities at the beginning and end of the experimental period. Because the soil texture of loamy sand was coarser than clay loam, the bulk density of loamy sand soil was higher than clay loamy soil. Adding zeolite to the soil caused an increase in the bulk density at the beginning of the experiment. Nonetheless, irrigation of the soil with the leachate of organic fertilizer factory significantly decreased the bulk density in all treatments ($p < 0.05$) at the end of the study period (Table 5). This result is consistent with previous studies (Chandrasekaran and Rajkannan 2003; Bhardwaj et al. 2007; Malla et al. 2007).

Table 5: Bulk density at the beginning and at the end of the experimental period

		A0	A5	A10	B0	B5	B10
Bulk density (g·cm3)	Beginning	1.88a*	1.94a	1.91a	1.50a	1.60a	1.64a
	Ending	1.68b	1.55b	1.65b	1.25b	1.20b	1.10b

*Values in any column followed by the same letter do not differ at the 0.05 significance level.

Mirzaei *et al.*

Mirzaei *et al. International Journal of Recycling of Organic Waste in Agriculture* 2013 **2**:20, doi: 10.1186/2251-7715-2-20

Saturated Hydraulic Conductivity (K_{sat})

The high K_{sat} in loamy sand soil was due to its coarser texture as compared to clay loamy soil (Table 6). Fine particles are a major cause of reduced soil K_{sat} in surface soils, particularly when irrigated with sodic waters with low electrolyte concentration Felhendler et al. (1974).

Table 6: Saturated hydraulic conductivity (m/day) for all treatments before and after application of leachate

Zeolite application	Loamy sand			Clay loam		
	0	5	10	0	5	10
Before start of experiment	3.25a*	3.17 a	2.88 a	1.73a	1.58a	1.44a
End of the experiment	3.25a	3.17 a	2.88 a	0.36b	0.29b	0.14b

*Values in any column followed by the same letter do not differ at the 0.05 significance level.

Mirzaei *et al.*

Mirzaei *et al. International Journal of Recycling of Organic Waste in Agriculture* 2013 2:20, doi: 10.1186/2251-7715-2-20

Along with increasing zeolite content in clay loamy soil treatments, K_{sat} decreased. Zeolite particle size distribution (0.4 to 2 mm) also affected K_{sat}. For loamy sand soils, application of zeolite effectively reduced soil hydraulic conductivity. Lin et al. (1998) also reported in their research that application of zeolite in sandy and loamy soils effectively reduced the average particle size and consequently lowered hydraulic conductivity.

However, it is inferred that the existence of a large amount of sodium in the influent solution causes enhancement of the sodium adsorption ratio (SAR) that consequently causes dispersion of the soil matrix. Clay dispersion in the clay loamy soil and high total suspended solids (TSS) are the major causes of hydraulic conductivity reduction. Previous studies have also mentioned a significant decrease in soil hydraulic conductivity after the application of effluent (Magesan et al. 1999; Tarchitzky et al. 1999; Halliwell et al. 2001; Levy et al. 2005; Francisca and Glatstein 2010).

Infiltration

Domestic and industrial wastewater does have usually high amount of sodium and this causes a reduction of water infiltration through the soil over the time. Reduced water infiltration inhibits soil profile's leaching and further concentration of sodium and other toxic ions Vogeler (2009).

As illustrated in Figure 6, infiltration decreased by adding zeolite to the soil. This is due to the tininess of the zeolite compared to the soil particles. These zeolite small particles can fill spaces of the soil particles and clog the pores of the soil. In loamy sand soil, no significant difference was revealed ($p < 0.05$) between amount of accumulated infiltration at the beginning and at the end of experimental period. Oppositely, in the clay loamy soil, soil infiltration significantly decreased along with increasing zeolite content at the beginning and at the end of the experimental period (data are not shown). As illustrated in Figure

6, soils with 10% zeolite (A10 and B10) possess the least amount of infiltration and the lowest hydraulic conductivity.

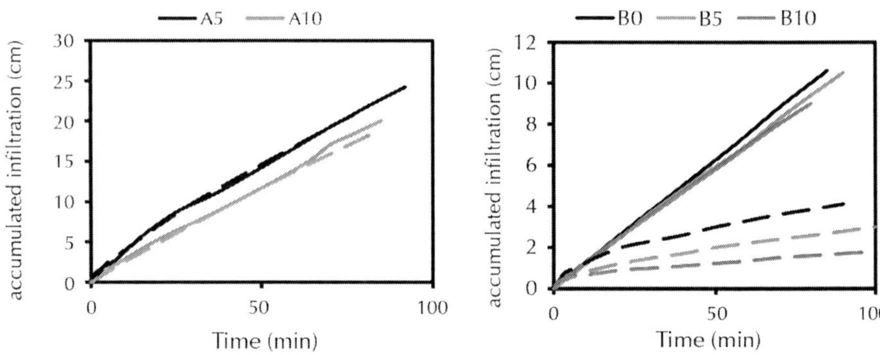

Figure 6: Accumulated infiltration at the beginning (Solid Line) and end (Dash Line) of experimental period.

Furthermore, high values of TSS and SAR of the leachate can reduce soil permeability. Lado et al. (2005) showed that using longer than 10 years of effluent for irrigation would cause abatement of infiltration. Deletic and Fletcher (2006) demonstrated that clogging of the soil due to the deposition of sediment may cause a decrease in infiltration. Some studies have reported long-term and constant use of water with a high adjusted SAR value, which would lead to a reduction of the soil infiltration and permeability (Wallach et al. 2005; Graber et al. 2006).

CONCLUSIONS

In sum, we attempted to investigate the effect of land treatment and zeolite on reduction of heavy metal concentrations in leachate. Most outstanding results of this research could be listed as below:

1. Land treatment plays a significant role in the elimination of heavy metals from leachate.
2. Soil texture has a fundamental role in the elimination process.
3. In loamy sand soil, different levels of zeolite had significant effects on the reduction of heavy metal concentrations in the effluent.
4. Irrigation of the soil with leachate reduced bulk density and infiltration and saturated hydraulic conductivity of the soil due to

high SAR and TSS of the leachate. Last but not least, zeolite had a damaging effect on mentioned processes because of its role in the release of Na and its effect on physical properties of the soil.

AUTHORS' CONTRIBUTIONS

SMJM collected the experimental data. SHT, MH and PN carried out the supervision on the data analysis. SEH helped in laboratory analysis.

ACKNOWLEDGMENTS

Researchers would like to thank the officials of the University of Sharekord, Islamic Azad University, Khorasgan Branch and Isfahan Compost Factory for their financial supports of this research project.

REFERENCES

1. Ackley MW, Yang RT (1991) Diffusion in ion-exchanged clinoptilolites. AICHE J 37(11):1645-1656
2. Ahmed OH, Sumalatha G, Nik-Muhamad AM (2010) Use of zeolite in maize (Zea mays) cultivation on nitrogen, potassium and phosphorus uptake and use efficiency. Int J Phys Sci 5(15):2393-2401
3. APHA (1995) Standard method for the examination of water and wastewater. Washington, D.C: American Public Health Association. 1566
4. Bhardwaj AK, Goldstein D, Azenkot A, Levy GJ (2007) Irrigation with treated wastewater under two different irrigation methods: effects on hydraulic conductivity of a clay soil. Geoderma 140(1–2):199-206
5. Bittell JE, Miller RJ (1974) Lead, cadmium, and calcium selectivity coefficients on montmorillonite, illite and kaolinite. J Environ Qual 3(3):250-253
6. Brady NC, Weil RR (2008) The nature and properties of soils. Prentice Hall/Pearson Education. 965 pages

7. Breck DW (1974) Zeolite molecular sieves. New York: Wiley.
8. Chandrasekaran N, Rajkannan B (2003) Continuous application of sewage effluent on soil physical properties. Madras Agri J 90:4-6
9. Deletic A, Fletcher TD (2006) Performance of grass filters used for stormwater treatment—a field and modelling study. J Hydrol 317:261-275
10. Echeverria JC, Morera MT, Mazkiaran C, Garrido JJ (1998) Competitive sorption of heavy metal by soils. Isotherms and fractional factorial experiments. Environ Pollut 101(2):275-284
11. Erdem E, Karapinar N, Donat R (2004) The removal of heavy metal cations by natural zeolites. J Colloid Interface Sci 280(2):309-314
12. Felhendler R, Schainberg I, Frenkel H (1974) Dispersion and the hydraulic conductivity of soils in mixed solutions. In: Transactions of the 10th international congress of soil science. (Moscow). Nauka Pub. House, Moscow, pp: 103–112
13. Francisca FM, Glatstein DA (2010) Long term hydraulic conductivity of compacted soils permeated with landfill leachate. Appl Clay Sci 49:187-193
14. Graber ER, Ben-Arie O, Wallach R (2006) Effect of sample disturbance on soil water repellency determination in sandy soils. Geoderma 136(1–2):11-19
15. Halliwell DJ, Barlow KM, Nash DM (2001) A review of the effects of wastewater sodium on soil physical properties and their implications for irrigation systems. Aust J Soil Res 39(6):1259-1267
16. Hatt BE, Fletcher TD, Deletic A (2007) Treatment performance of gravel filter media: implications for design and application of stormwater infiltration systems. Water Res 41:2513-2524
17. Inglezakis VJ, Loizidou MD, Grigoropoulou HP (2002) Equilibrium and kinetic ion exchange studies of Pb^{2+}, Cr^{3+}, Fe^{3+} and Cu^{2+} on natural clinoptilolite. Water Res 36(11):2784-2792
18. Iskander AL, Khald EM, Sheta AS (2011) Zinc and manganese sorption behavior by natural zeolite and bentonite. Ann Agric Sci 56:43-48

19. Kalbassi M, Gandomkar A (1997) Effect of garbage leachate on yield and chemical composition of corn and the effect of leachate residual on soil characteristics. JWSS 1(2):41-52
20. Klute A (1986) Methods of soil analysis. Part 1, Physical and mineralogical methods. Madison, Ws, USA: Soill Sci. Sco. Am. J. p 1189
21. Lado M, Ben-Hur M, Assouline S (2005) Effects of effluent irrigation on seal formation, infiltration, and soil loss during rainfall. Soil Sci Soc Am J 69(5):1432-1439
22. Leppert D (1990) Heavy metal sorption with clinoptilolite zeolite: alternatives for treating contaminated soil and water. Min Eng 42(6):604-608
23. Levy GJ, Goldstein D, Mamedov AI (2005) Saturated hydraulic conductivity of semiarid soils: combined effects of salinity, sodicity, and rate of wetting. Soil Sci Soc Am J 69(3):653-662
24. Li J, Hu J, Sheng G, Zhao G, Huang Q (2009) Effect of pH, ionic strength, foreign ions and temperature on the adsorption of Cu(II) from aqueous solution to GMZ bentonite. Colloid Surface A 349(1–3):195-201
25. Lin CF, Lo SS, Lin HY, Lee YC (1998) Stabilization of cadmium contaminated soils using synthesized zeolite. J Hazard Mater 60(3):217-226
26. Magesan GN, Williamson JC, Sparling GP, Schipper LA, Lloyd-Jones AR (1999) Hydraulic conductivity in soils irrigated with wastewaters of differing strengths: field and laboratory studies. Aust J Soil Res 37(2):391-402
27. Mahabadi AA, Hajabbasi MA, Khademi H, Kazemian H (2007) Soil cadmium stabilization using an Iranian natural zeolite. Geoderma 137(3–4):388-393
28. Malla R, Tanaka Y, Mori K, Totawat KL (2007) Effect of short-term sewage irrigation on chemical build up in soils and vegetables. The Agricultural Engineering International. J of CIGR: Manuscript LW 07006. 1(IX):1-14
29. Mondale KD, Carland RM, Aplan FF (1995) The comparative ion exchange capacities of natural sedimentary and synthetic zeolites. Miner Eng 8(4–5):535-548

30. Noori M, Ahmadi A, Zendehdel M (2007) Comparative study between using natural and synthetic zeolites for the improvement of soil salinity and crop yield. Toxicol Environ Chem 89(1–4):233-241
31. Oste LA, Lexmond TM, Van Riemsdijk WH (2002) Metal immobilization in soils using synthetic zeolites. J Environ Qual 31(3):813-821
32. Ouki SK, Kavannagh M (1999) Treatment of metals-contaminated wastewaters by use of natural zeolites. Water Sci Technol 39(10–11):115-122
33. Perez-Caballero R, Gil J, Benitez C, Gonzalez JL (2008) The effect of adding zeolite to soils in order to improve the N-K nutrition of olive trees. Preliminary results. Am J Agric Biol Sci 3(1):321-324
34. Pietri JCA, Brookes PC (2008) Nitrogen mineralisation along a pH gradient of a silty loam UK soil. Soil Biol Biochemistry 40(3):797-802
35. SAS (1987) Statistical Analysis System (SAS. Cary, NC 27513–2414, USA: SAS Institute Inc.
36. Shanableh A, Kharabsheh A (1996) Stabilization of Cd, Ni and Pb in soil using natural zeolite. J Hazard Mater 45(2–3):207-217
37. Sheng GD, Wang SW, Hu J, Lu Y, Li JX, Dong YH, Wang XK (2009) Adsorption of Pb(II) on diatomite as affected via aqueous solution chemistry and temperature. Colloid Surface A 339(1–3):159-166
38. Shi W, Shao H, Li H, Shao M, Du S (2009) Progress in the remediation of hazardous heavy metal-polluted soils by natural zeolite. J Hazard Mater 170(1):1-6
39. Tabatabaei SH, Najafi P, Mirzaei SMJ, Nazem Z, Heidarpour M, Hajrasoliha S, Afuni M, BeigiHarchegani H, Landi E, Akasheh L, Zamanian M, Barani M, Amini H (2012) Compost' leachate recycling through land treatment and application of natural Zeolite. Int J Recycling Organic Waste Agri 1(2):1-7
40. Tarchitzky J, Golobati Y, Keren R, Chen Y (1999) Wastewater effects on montmorillonite suspensions and hydraulic properties of sandy soils. Soil Sci Soc Am J 63(3):554-560
41. Thawale PR, Juwarkar AA, Singh SK (2006) Resource conservation through land treatment of municipal wastewater. Curr Sci 90(5):704-711

42. Turan NG (2008) The effects of natural zeolite on salinity level of poultry litter compost. Bioresour Technol 99(7):2097-2101
43. Vogeler I (2009) Effect of long-term wastewater application on physical soil properties. Water Air Soil Pollut 196(1–4):385-392
44. Wallach R, Ben-Arie O, Graber ER (2005) Soil water repellency induced by long-term irrigation with treated sewage effluent. J Environ Qual 34(5):1910-1920
45. Wang S, Hu J, Li J, Dong Y (2009) Influence of pH, soil humic/fulvic acid, ionic strength, foreign ions and addition sequences on adsorption of Pb(II) onto GMZ bentonite. J Hazard Mater 167(1–3):44-51
46. Yang X, Yang S, Yang S, Hu J, Tan X, Wang X (2011) Effect of pH, ionic strength and temperature on sorption of Pb(II) on NKF-6 zeolite studied by batch technique. Chem Eng J 168(1):86-93
47. Zhang H, Yu XJ, Chen L, Geng JQ (2010) Investigation of radionuclide Ni-63(II) sorption on ZSM-5 zeolite. J Radioanal Nucl Chem 286(1):249-258

Chapter 8

Assessment of Soil Pollution with Heavy Metals in Romania

Anca Maria Moldoveanu[1,2]

[1]University of Medicine and Pharmacy, "Carol Davila" Bucharest, Romania

[2]National Institute of Public Health, Bucharest, Romania

INTRODUCTION

Human activities, industry, erosion and the continuous spread of urbanization can induce soil pollution in various ways. The pollution of soil can be dangerous for human health because the toxic substances can enter the crops and the ground water. The soil, due to its properties and structure, plays the role of a filter that can retain and be a deposit for toxic substances. The most frequent contaminants of soil in Europe are heavy metals and mineral oil. The group 'heavy metals' for the purpose of discussing health risks or impacts generally includes: Arsenic (As), Lead (Pb), Cadmium (Cd), Chromium (Cr) (although only the form Cr(VI) is toxic), Copper (Cu), Mercury (Hg), Nickel (Ni), Zinc (Zn).

Several of these elements are necessary for human health and are beneficial when taken into the body in foods or as supplements at appropriate, low levels. Conversely, cadmium, lead and mercury have no known biological function and are toxic to humans.

The sources of heavy metals that pollute the soil can be human activities like metalliferous mining (As, Cd, Cu, Ni, Pb, Zn), smelters (As, Cd, Pb), metallurgy, electronic industry (where metals are used in batteries, semiconductors, circuits), rolling (Ni, Cd, Pb, Hg, Se), dyes and paints industry (Pb, Cr, As, Se, Mo, Cd, Co, Ba, Zn), plastics industry (Cd, Zn, Pb, Sn are used as polymer stabilizers), chemical industry (using Pb, Ni, Nb, Hg, Pt, Ru as electrode catalysts), wood industry (As, Cr and Cu). In the vicinity of furniture factories and wood-processing, these elements were often identified as soil and water pollutants [1, 2, 3, 4, 5, 6, 7, 8, 9].

Also, by storing municipal waste, special and hazardous waste, the soil may be contaminated with various heavy metals.

The combustion of fossil fuels leads to the presence of Cd, Zn, As, Se, Cu, Mn, V, in the ashes and particulates of combustion. Some heavy metals (Se, Te, Pb, Mo, Li) are added to fuels and lubricants to improve their properties. [1, 2, 3, 4, 5, 6, 7, 8, 9].

Corrosion of metals contained in various materials (Cu, Pb pipes and roofs, Cr, Ni, Co in stainless steel, Cr, Pb painting) can cause soil pollution. Agro-livestock activities can also cause soil pollution (As, Cu, Zn can be added to feed pigs and poultry, Cd, As, Pb, Mn, Cu, Zn can be present in phosphate fertilizers and pesticides). [1, 2, 3, 4, 5, 6, 7, 8, 9].

Pollution by heavy metals and many organic contaminants is practically irreversible (European Commission, 2012).

Excess heavy metal accumulation in soils is toxic to humans and other animals.

Exposure to heavy metals is normally chronic (exposure over a longer period of time), due to food chain transfer. Acute (immediate) poisoning from heavy metals is rare, but possible, through ingestion or dermal contact. Chronic problems associated with long-term heavy metal exposures are [10, 11, 12,13, 14, 15, 16, 17, 18, 19] for:

- Lead [20-60] – is well known to be toxic and its effects have been more extensively reviewed than the effects of other trace metals. Lead can cause serious injury to the brain, nervous system, red blood cells, and kidneys. Exposure to lead can result in a wide range of biological effects depending on the level and duration of exposure. Various effects occur over a broad range of doses, with the developing young and infants being more sensitive than adults. Lead poisoning severe enough to cause evident illness is now very rare. Lead can be harmful after uptake from food, air, water or contaminated dust. Lead is a particularly dangerous chemical, as it can accumulate in individual organisms, as well as entire food chains.
- Cadmium – affects kidney, liver, and GI tract [28, 30, 61, 62, 63].
- Arsenic – is associated with skin damage, increased risk of cancer and problems with circulatory circulatory system, affects kidneys and the central nervous system [64, 65, 66, 67, 68, 69].
- Chromium is associated with allergic dermatitis in humans [70, 71].
- Mercury is associated with kidney damage [72, 73, 74, 75].
- The most common problem caused by cationic metals (metallic elements whose forms in soil are positively charged cations e.g., $Pb2+$) are mercury, cadmium, lead, nickel, copper, zinc, chromium and manganese. The most common anionic compounds (elements whose forms in soil are combined with oxygen and are negatively charged e.g., $MoO4^{2-}$) are arsenic, molybdenum, selenium, and boron.

The Aim of the Study

This study tried to assess the level of soil contamination with heavy metals (lead, cadmium, mercury, manganese, nickel, total chromium, zinc, cobalt, copper and arsenic) in 34 counties in Romania. No previous data exists as a reference level of soil pollution.

MATERIAL AND METHODOLOGY

Sampling was done in 2012, in the fall (September – October). Soil samples were taken from max 20 cm deep, from areas around children's units like nurseries, kindergartens, schools or playgrounds for children. The samples were taken in county capital cities. Also, soil samples were taken from areas of land used for agricultural purposes like farmlands and gardens.

For metal analysis, 0.2 g of ground soil were digested with aqua regia using microwave digestion (EN 13346 :02) followed by filtration and adjusting the volume in a volumetric flask. Concentrations of Pb, Cd, Cu, Cr were measured using the electrothermal atomic absorption spectrometry (ISO 11464:2006, SR ISO 11047/1999, EPA 6020). Concentrations of Zn, Mn, Co, Ni, Cr were measured using flame AAS (atomic absorption spectrometry) and concentrations of Hg and As were measured using atomic absorption spectrometry with cold vapour (HVG-AAS) and SR EN 1485/2007. [76, 77]

Despite the importance of soil pollution for our society, there is no EU legislation specifically targeting the protection of soil [78, 79]. A regulation regarding the soil pollution exists in Romania (Order nr. 756, November, the 3th, 1997). In this Order [80], the following definitions are of importance: environmental impact, types of land uses that require a certain soil quality, soil characterized by a maximum acceptable level of pollutants - sensitive and less sensitive soil, remediation targets, potentially significant pollution, significant pollution, threshold alert – (concentrations of pollutants in air, water, soil or emissions / discharges, which are designed to alert the competent authorities of a potential impact on the environment and determines the triggering of additional monitoring and / or reduction of concentrations of pollutants in emissions / evacuations), intervention threshold, reference sample [80].

Table 1: Reference values for traces of heavy metals in soil - Heavy metals (mg/kg dry substance). Values are stipulated in Order 756/1997 for the approval of the Regulation on the assessment of environmental pollution [80]

Heavy Metals	Reference threshold for sensitive soil	Alert thresholds - sensitive soil	Alert thresholds - soil less sensitive	Action levels - soil sensitive	Action levels – soil less sensitive
Arsenic (As)	5	15	25	25	50
Cadmium (Cd)	1	3	5	5	10
Cobalt (Co)	15	30	100	50	250
Chromium (Cr) : Total Chromium total Hexavalent Chromium	30 1	100 4	300 10	300 10	600 20
Copper (Cu)	20	100	250	200	500
Manganese (Mn)	900	1500	2000	2500	4000
Mercury (Hg)	0.1	1	4	2	10
Nickel (Ni)	20	75	200	150	500
Lead(Pb)	20	50	250	100	1000
Zinc (Zn)	100	300	700	600	1500

RESULTS AND DISCUSSIONS

Cadmium

Cadmium in the soil registered levels that ranged from 0 and 0.86 mg/kg, in the urban areas and in the rural areas. Compared to the threshold value it can be seen that the levels of cadmium in soil, in both urban and rural areas, have not exceeded the threshold value (1mg/kg). It can be seen that the level of cadmium in soil is slightly higher in rural areas compared to urban areas.

Lead

Lead in soil, registered levels that ranged from 0 and 96.11 mg/kg in the urban areas and from 0 to 147.67 mg/kg in rural areas. It can be seen that, both in the urban and rural areas, the lead levels in soil exceeded the threshold value (20 mg/kg). The alert threshold for sensitive soil (50 mg/kg) was exceeded in both the urban and rural areas.

Figure 1: Lead levels in soil in urban areas.

Figure 2: Lead levels in soil in rural areas.

Comparing the two charts it can be seen that, in the same area (Constanța county), the level of lead in soil is over the alert threshold in the urban and in the rural area. In other two counties (Buzău and Ialomița) the levels of lead in soil are exceeding the threshold limit both in the urban area and in the rural area. There are other two counties beside Constanta, where the levels of lead are exceeding the alert threshold in urban areas (Olt and Sibiu). In rural areas it can be seen that the number of counties where lead levels in soil exceed the threshold is higher than in urban areas.

Copper

Copper in soil registered levels that ranged from 0 and 128.5 mg/kg in the urban areas and from 0 to 80.43 mg/kg in rural areas. It can be seen that the levels identified in soil exceeded, both in the urban and rural areas, the threshold value (20 mg/kg). Also, the alert threshold for sensitive soil (100 mg/kg) was exceeded in a single urban area. In rural areas a slightly higher level of copper in the soil was observed.

Figure 3: Copper levels in soil in urban areas.

Figure 4: Copper levels in soil in rural areas.

The alert threshold was exceeded in a single county in urban areas. The number of counties where was found levels of copper in soil, in rural areas, that exceeded the threshold was double in comparison with urban areas.

Nickel

Nickel in soil registered levels that ranged from 0 and 44 mg/kg in the urban areas and from 0 to 41.6 mg/kg in rural areas. The levels of nickel found in soil samples exceeded the threshold value (20 mg/kg), in the urban and rural areas. Nickel recorded slightly higher levels in the soil of rural areas, but, also, were found peaks in urban soil. The alert threshold for sensitive soil (75 mg/kg) was not exceeded in either urban or rural area.

Figure 5: Nickel levels in soil in urban areas.

Figure 6: Nickel levels in soil in rural areas.

As it can be seen in the charts above, the soil pollution with nickel is similar in the urban and rural areas, regarding the number of counties affected.

Manganese

Manganese in soil registered levels that ranged from 0 and 1256.57 mg/kg in the urban areas and in rural areas the levels ranged from 0 to 1072. 62 mg/kg. Compared to the threshold value, the levels of manganese found in soil exceeded the threshold, both in the urban and rural areas (900 mg/kg). Alert threshold (1500 mg/kg) was not exceeded in neither rural areas nor urban areas.

FIGURE 7: Manganese levels in soil in urban areas.

Figure 8: Manganese levels in soil in rural areas.

The threshold was exceeded, both in urban and rural areas, in only one county in the middle of the country (Mureş).

Total Chromium

Concerning the levels of total chromium in soil, values are similar, in urban and rural areas. The levels found in soil were between 0 and

40.81 mg/kg in urban areas and between 0 and 57.23 mg/kg in rural areas. Only the threshold (30 mg/kg) was exceeded in the urban and rural areas and not the alert threshold (100 mg/kg).

Figure 9: Total chromium levels in soil in urban areas.

Figure 10: Total chromium levels in soil in rural areas.

The total chromium levels registered in soil exceeded the threshold in a small number of counties. Only in one county (Mureş), levels of chromium exceeding the threshold were found in both the urban and rural areas.

Cobalt

Cobalt levels were higher in rural areas compared to urban areas. The highest value was found in the soil of a farmland in Covasna county, where the level of cobalt was equal to 20.23 mg/kg, exceeding the threshold value (15 mg/kg) but not the alert threshold (30 mg/kg).

Figure 11: Total cobalt levels in soil in urban areas.

Figure 12: Total cobalt levels in soil in rural areas.

From the charts, it can be seen that it is only one hot spot where the levels of cobalt in soil exceeded the threshold.

Zinc levels

Zinc levels ranged between 0 and 275.35 mg/kg in the soil of urban areas and between 0 – 131.79 mg/kg in the soil of rural areas. The levels recorded were similar in urban and rural areas. One single higher value was found in Sibiu county.

Figure 13: Zinc levels in soil in urban areas.

Figure 14: Zinc levels in soil in rural areas.

A similar situation with zinc pollution of the soil was found in the urban and in the rural areas of Romania as number of counties affected.

Mercury

Mercury levels in the soil ranged between 0 and 0.265 mg/kg in the soil of urban areas and between 0 – 0.305 mg/kg in the soil of rural areas. The threshold value is equal to 0.1 mg/kg and was exceeded in urban and rural areas, but the alert threshold is equal to 1 mg/kg and was not exceeded in neither urban nor in rural areas. Special mention must be made for district Ferneziu of the town Baia Mare where there is a historical pollution of the environment. The level of mercury found in this area was equal to 0.265 mg/kg.

Figure 15: Mercury levels in soil in urban areas.

Figure 16: Mercury levels in soil in rural areas.

For the situation of soil pollution with mercury it must be stressed that, in the area of Maramureș there is a historical pollution due to a non-ferrous industry existing for a long period of time.

Arsenic

Arsenic levels in soil were between 0 and 8.22 mg/kg in urban areas and 0 and 9.59 mg/kg soil. Arsenic concentrations were higher in rural areas compared to urban soil. The reference threshold is equal to 5 mg/kg and the alert threshold is equal to 15 mg/kg. The threshold value was exceeded but the not the alert threshold.

Figure 17: Arsenic levels in soil in urban areas.

Figure 18: Arsenic levels in soil in rural areas.

Comparing the two maps, it can be seen that, levels of pollution exceeding the threshold were found in the Sibiu and Mureş County, in the middle of the country, in both urban and rural areas.

CONCLUSIONS

This study is the first study that tried to assess the situation of soil pollution with heavy metals in Romania.

After assessment of the soil contamination with heavy metals in 34 counties in Romania, it can be concluded that for all investigated metals except cadmium the threshold values were exceeded.

Alert thresholds were exceeded for lead and copper, both in urban and rural areas, especially in two counties: Maramureș county and Sibiu county. Areas like Maramureș county or Sibiu county have a historic pollution due to the industry present since 1844 in Baia Mare, Maramures county and 1936 în Copșa Mică, Sibiu county.

A national screening program was performed in the years 2008-2010 in Maramureș county (Baia Mare city and especially Ferneziu district) to see the impact on human health in connection with lead pollution. In the years 2008 and 2009 a group of 91 children was investigated [81, 82].

The results showed the following situation:

Table 2: Lead levels in blood in city Baia Mare, Maramureș county in 2008 and 2009

Year of study	Lead in blood up to 10 μg/dl	Lead in blood up to 20 μg/dl	Lead in blood up to 30 μg/dl	Lead in blood up to 35 μg/dl	Lead in blood > 35 μg/dl
2008 – % of investigated children	95.4	97.8	97.8	97.8	2.2

2009 – % of investigated children	58.70	71.74	84.06	89.13	10.87

A special situation had the children living near the factory that processes lead in the district Ferneziu in Baia Mare county.

During the year 2010, the study group was made of 94 pregnant women and 94 new born babies. The blood was taken from the new born cord in the case of children [83].

Table 3: Lead levels in blood in the group of pregnant women, in city Baia Mare, Maramureș county, in 2010.

Level of lead in blood of pregnant women	0-10 µg/dl	10-20 µg/dl	20-30 µg/dl	> 35 µg/dl
% of the study group	94.7	1.06	2.12	2.12

Table 4: Lead levels in blood in the cord of new born babies, in city Baia Mare, Maramureș county, in 2010

Level of lead in blood of cord of new born babies	0-10 µg/dl	10-20 µg/dl	20-30 µg/dl	> 35 µg/dl
% of the study group	93.6	1.06	-	5.34

For copper, alert thresholds was exceeded, both in urban and rural areas. In Suceava county, copper mining was performed since 1922. In this area, there were taken rehabilitation measures in the period between1999-2005. Still, some of the effects of the ancient pollution situation can be seen today.

All the other investigated metals in the soil, both in residential areas and agricultural areas (which both can be classified as sensitive lands), do not exceed the alert threshold for sensitive lands.

For cadmium, nickel, manganese, cobalt and arsenic higher levels were recorded in the soil of rural areas than in the soil of urban areas.

For total chromium, zinc and mercury similar levels were recorded in the soil of urban and rural areas.

It must be stressed that in the areas where the alert thresholds were exceeded, pollution with impact on the environment can be present. The situation can be labelled as "potentially significant pollution". In this case, the competent authorities can take action to reduce the pollutants emissions in the air or to take measures to diminish the possibilities to contaminate the soil directly or through water.

REFERENCES

1. Raymond A.Wuana and Felix E. Okieimen, HeavyMetals in Contaminated Soils: A Review of Sources, Chemistry, Risks and Best Available Strategies for Remediation; International Scholarly Research Network ISRN Ecology, Volume 2011, Article ID 402647, 20 pages, doi:10.5402/2011/402647,
2. Jintao Liang , Cuicui Chen, Xiuli Song, Yulan Han, Zhenhai Liang , Assessment of Heavy Metal Pollution in Soil and Plants from Dunhua Sewage Irrigation Area, Int. J. Electrochem. Sci., 6 (2011) 5314 – 5324, www.electrochemsci.org,
3. Heavy Metal Soil Contamination, Soil Quality – Urban Technical Note No. 3, United States Department of Agriculture, September, 2000,
4. EPA. Soil Screening Guidance: User's Guide. Publication 9355.4, July 23th, 1996
5. European Commission, Science for Environment Report, In-Depth Report, Soil Contamination: Impacts on Human Health, September 2013, Issue 5. http://ec.europa.eu/environment.
6. EUGRIS *Portal for land and soil management in Europe* [online]. Available at: www.eugris.info
7. European Commission (2012) *The State of Soil in Europe – A contribution of the JRC to the European Environment Agency's State and Outlook Report–SOER 2010*. Luxembourg: Publications Office of the European Union. Available at: http://ec.europa.eu/dgs/jrc/downloads/ jrc_reference_report_2012_02_soil.pdf

8. Rodriguez Lado, L., Hengl, T., Reuter, H.I., (2008) Heavy metals in European soils: a geostatistical analysis of the FOREGS Geochemical database.*Geoderma* 148: 189-199.
9. Huber & Prokop (2012), Progress in the management of contaminated sites. EIONET Workshop on Soil, Ispra, 10- 12 December 2012.
10. D. R. Baldwin and W. J. Marshall, "Heavy metal poisoning and its laboratory investigation, " *Annals of Clinical Biochemistry*, vol. 36, no. 3, pp. 267–300, 1999.
11. WHO/UNECE (2006): *Health risks of heavy metals from long-rangetransboundary air pollution*. Draft of May 2006. Copenhagen: WorldHealth Organization Regional Office for Europe and Geneva: UnitedNations Economic Commission for Europe (UNECE). Available at:www.euro.who.int/__data/assets/pdf_file/0007/78649/E91044.pdfWHO (2013) *Ten chemicals of major health concern*. [online]. Retrievedfrom: www.who.int/ipcs/assessment/public_health/chemicals_phc/en/index.html,
12. Brevik, E.C. & Burgess, L.C. eds (2013) *Soils and Human Health*. Boca Raton: CRC Press.
13. Kibble, A. and Russell, D. (2010) *Contaminated Land and Health*. In Maynard, R.L. *et al*. (eds) Environmental Medicine. Boca Raton, Florida: CRC Press. Pp: 565–573
14. Meerschman, E., Cockx, L., Islam, M.M., Meeuws, F., and Van Meirvenne, M. (2011) Geostatistical Assessment of the Impact of World War I on the Spatial Occurrence of Soil Heavy Metals. *Ambio*.40(4): 417–424.
15. Navarro, M.C., Perez-Sirvent, C., Martinez-Sanchez, M.J., Vidal, J., Tovar, P.J., Bech, J. (2008) Abandoned mine sites as a source of contamination by heavy metals: A case study in a semi-arid zone. *Journal of Geochemical Exploration* 96: 183–193
16. Environment Agency (England) (2009) *Human health toxicological assessment of contaminants in soil*. Science report - Final SC050021/ SR2. Bristol: Environment Agency. Available at: www.environmentagency. gov.uk/static/documents/Research/TOX_guidance_report_-_ final.pdf
17. Oliver, M.A. (1997) Soil and human health: a review. *European Journal of Soil Science*.48: 573-592

18. Morgan, R. (2013) *Soil, Heavy Metals, and Human Health*. In Brevik, E.C. & Burgess, L.C. (2013) Soils and Human Health. Boca Raton. FL: CRC Press, pp. 59-80.
19. Gil, F., Capitan-Vallvey, L.F., De Santiago, E., Ballesta, J., Pla, A., Hernandez, A.F., Gutierrez-Bedmar, M., Fernandez-Crehuet, J., Gomez, J., Lopez-Guarnido, O., Rodrigo, L., Villanueva, E. (2006) Heavy metal concentrations in the general population of Andalusia, South of Spain A comparison with the population within the area of influence of Aznalcollar mine spill (SW Spain). *Science of the Total Environment*. 372 (2006): 49-57.
20. WHO Europe (2009) *Levels of lead in children's blood fact sheet 4.5 (RPG4_Chem_Ex1)* Copenhagen: World Health OrganizationRegional Office for Europe. Available at: www.euro.who.int/__data/ assets/pdf_file/0003/97050/4.5.-Levels-of-lead-in-childrens-blood- EDITING_layouted.pdf
21. Canfield, R.L., Henderson, C.R. Jr, Cory-Slechta, D.A., Cox, C., Jusko, T.A., Lanphear, B.P. (2003) Intellectual impairment in children with blood concentrations below 10 µg per deciliter. *New England Journal of Medicine*. 348:1517–1526.
22. Chen, A., Dietrich, K.N., Ware, J.H., Radcliffe, J., Rogan, W.J. (2005) IQ and blood lead from 2 to 7 years of age: are the effects in older children the residual of high blood lead concentrations in 2-year-olds? *Environmental Health Perspectives*. 113:597–601.
23. Jakubowski, M., Trzcinka-Ochocka. M., Raźniewska, G., Christensen, J.M., Starek, A. (1996) Blood lead in the general population in Poland.*International Archives of Occupational and Environmental Health*. 68:193–198.
24. Mielke, H.W. & Reagan, P.L. (1998) Soil is an important pathway of human lead exposure. *Environmental Health Perspectives*. 106 (Suppl 1): 217–229.
25. Tong, S., von Schirnding, Y.E., Prapamontol, T. (2000) Environmental lead exposure: a public health problem of global dimensions. *Bulletin of the World Health Organization*. 78:1068–1077.
26. Ducoffre, G. et al. Lowering time trend of blood lead levels in Belgium since 1978. *Environmental research*, 51: 25–34 (1990).
27. *Inorganic lead*. Geneva, World Health Organization, 1995 (Environmental Health Criteria, No. 165).

28. Elinder, C.G. et al. Lead and cadmium levels in blood samples from the general population of Sweden. *Environmental research*, 30: 233–253 (1983).
29. Brunekreef, B. The relationship between air lead and blood lead in children: a critical review. *Science of the total environment*, 38: 79–123 (1984).
30. Ewers, U. et al. Contribution of lead and cadmium in dust fall to blood lead and blood cadmium in children and adults living in two nonferrous smelter areas of West Germany.*In*: Lekkas, T.D., ed. *International Conference on Heavy Metals in the Environment, Athens, September 1985*. Edinburgh, CEP Consultants, 1985.
31. *Guidance manual for the integrated exposure uptake biokinetic model for lead in children*. Washington, DC, US Environmental Protection Agency, 1994 (EPA 540-R-93-081).
32. Ziegler, E.E. ET AL. Absorption and retention of lead by infants. *Pediatric research*, 12:29–34 (1978).
33. Brunekreef, B. *The relationship between environmental lead and blood lead in children: a study in environmental epidemiology*. Wageningen, Department of Environmental and Tropical Health, Agricultural University of Wageningen, 1985(Report 1985-211).
34. *Lead*. Geneva, World Health Organization, 1977 (Environmental Health Criteria, No. 3).
35. Barry, P.S.I. Concentrations of lead in the tissues of children. *British journal of industrial medicine*, 38: 61–71 (1981).
36. Buchet, J.P. et al. Mobilization of lead during pregnancy in rats. *International archives of occupational and environmental health*, 40: 33–36 (1977).
37. Silbergeld, E.K. Lead in bone: Implications for toxicology during pregnancy and lactation. *Environmental health perspectives*, 91: 63–70 (1991).
38. Silbergeld, E.K. et al. Lead and osteoporosis: mobilization of lead from bone in postmenopausal women. *Environmental research*, 47: 79–94 (1989).
39. Grandjean, P. et al. Reference intervals for trace elements in blood: significance of risk factors. *Scandinavian journal of clinical and laboratory investigation*, 52: 321–337 (1992).

40. Smith, M. et al. The effects of lead exposure on urban children: the Institute of Child Health/Southampton Study. *Developmental medicine and child neurology*,. 47 (Suppl):1–54 (1983).
41. Fergusson, D.M. et al. A longitudinal study of dentine lead levels, intelligence, school performance and behaviour. Part I Dentine lead levels and exposure to environmental risk factors. *Journal of child psychology and psychiatry*, 29: 781–792 (1988).
42. Baghurst, P. et al. Lifelong exposure to environmental lead and children's intelligence at age seven: the Port Pirie cohort study. *New England journal of medicine*, 327: 1269–1284 (1992).
43. Winneke, G. et al. Follow-up studies in lead-exposed children. *In*: Smith, M.A. et al., ed. *Lead exposure and child development, an international assessment*. Dordrecht, Kluwer Academic Publishers, 1989, pp. 260–270.
44. Altmann, L. et al. Impairment of long-term potentiation and learning following chronic lead exposure. *Toxicology letters*, 66: 105–112 (1993).
45. *Air quality criteria for lead*. Research Triangle Park, NC, US Environmental Protection Agency, 1986.Chapter 6.7 Lead Air Quality Guidelines - Second Edition,
46. *Air quality criteria for lead*. Research Triangle Park, NC, US Environmental Protection Agency, 2000.Chapter 6.7 Lead Air Quality Guidelines
47. *Toxicological profile for lead – draft for public comment*. Atlanta, GA, US Department of Health and Public Services, 1992.
48. Hernberg, S. & Nikkanen, J. Enzyme inhibition by lead under normal urbanconditions. *Lancet*, 1: 63–64 (1970)
49. Piomelli, S. et al. Threshold for lead damage to heme synthesis in urban children.*Proceedings of the National Academy of Sciences of the USA*, 79: 3335–3339 (1982).
50. Grandjean, P. et al. Delayed blood regeneration in lead exposure: an effect on reserve capacity. *American journal of public health*, 79: 1385–1388 (1989).
51. Piotrowski, J.K. & O'brien, B.J. *Analysis of the effects of lead in tissue upon human health, using dose–response relationships*. London, University of London Monitoring and Assessment Research Centre, 1980 (MARC Report No. 17).

52. Schwartz, J. et al. Lead-induced anemia: dose–response relationships and evidence for a threshold. *American journal of public health,* 80: 165–168 (1990).
53. Seppäläinen, A.M. et al. Subclinical neuropathy at "safe" levels of lead exposure.
54. *Archives of environmental health,* 30: 180–183 (1983).
55. Davis, J.M. and Svendsgaard, D.J. Nerve conduction velocity and lead: a critical review and meta-analysis. *In*: Johnson, B.L., ed. *Advances in neurobehavioural toxicology*. Chelsea, Lewis Publishers, 1990, pp. 353–376.
56. Stollery, B.T. et al. Short term prospective study of cognitive functioning in lead. *British journal of industrial medicine,* 48: 739–749 (1991).
57. Schwartz, J. Low-level lead exposure and children's IQ: A meta-analysis and search for a threshold. *Environmental research,* 65: 42–55 (1994).
58. Needleman, H. et al. Lead-associated intellectual deficit (letter). *Journal of the American Medical Association,* 306:367 (1982)
59. Staessen, J.A. et al. Impairment of renal function with increasing blood lead concentrations in the general population. *New England journal of medicine,* 327: 151–156 (1992).
60. Schwartz, J. & Otto, D.A. Blood lead, hearing thresholds and neurobehavioral development in children and youth. *Archives of environmental health,* 42: 153–160 (1987).
61. Schwartz, J. & Otto, D.A. Lead and minor hearing impairment. *Archives of environmental health,* 46: 300–305 (1991).
62. Pan, J., Plant, J.A. Voulvoulis, N., Oates, C.J. Ihlenfeld, C. (2010). Cadmium levels in Europe: implications for human health. *Environmental Geochemistry and Health.* 32 (1): 1-12
63. Schoeters, G., Den Hond, E., Zuurbier, M., Naginiene, R., Van Den Hazel, P., Stilianakis, N., Ronchetti, R., Koppe, J.G. (2006): Cadmium and children: Exposure and health effects. *Acta Paediatrica,* 95(Suppl.): 50-54.
64. UNEP (2010) *Final review of scientific information on cadmium.* Nairobi: United Nations Environment Programme. Available at: www.unep.org/hazardoussubstances/Portals/9/Lead_Cadmium/

docs/ Interim_reviews/UNEP_GC26_INF_11_Add_2_Final_ UNEP_ Cadmium_review_and_apppendix_Dec_2010.pdf,

65. WHO (2010a) *Exposure to Arsenic: A major public health concern*. Geneva: World Health Organization. Available at: www.who.int/ipcs/ features/arsenic.pdf,
66. Xu, J. and Thornton, I. (1985) Arsenic in garden soils and vegetable crops in Cornwall, England: Implications for human health. *Environmental Geochemistry and Health*. 7(4): 131-133.
67. Klinck, B., Palumbo, B., Cave, M. and Wragg, J. (2005) *Arsenic dispersal and bioaccessibility in mine contaminated soils: a case study from an abandoned arsenic mine in Devon, UK*. British Geological Survey Research Report RR/04/003 52pp. (ISBN 0 85272 483 7). Available at: http://nora.nerc.ac.uk/3681/1/ RR04003.pdf
68. Martinez-Sanchez, M.J., Martinez-Lopez, S., Martinez-Martinez, L.B., Perez-Sirvent, C. (2013) Importance of the oral arsenic bioaccessibility factor for characterising the risk associated with soil ingestion in a mining-influenced zone. *Journal of Environmental Management*. 116: 10-17.
69. Mitchell, P. and Barre, D. (1995) The nature and significance of public exposure to arsenic: a review of its relevance to South West England.*Environmental Geochemistry and Health*. 17(2):57-82.
70. Philipp, R., Hughes, A.O., Robertson, M.C. and Mitchell, T.F. (1984) *Soil levels of arsenic and malignant melanoma incidence*. In: Environmental Contamination, pp. 432–437. CEP Consultants, Edinburgh.
71. Chromium, Safety 101: Health Effects of Hexavalent Chromium, Minnesota Safety Council Minnesota OSHA at http://www.dli.mn.gov/OSHA/PDF/fact_hexchrom.pdf., 2006,
72. ATSDR – Chromium - ToxFAQsTM Internet address is http://www.atsdr.cdc.gov/toxfaqs/index.asp, October 2012,
73. UNEP (2013b) Global Mercury Assessment 2013: Sources, Emissions, Releases and Environmental Transport. Geneva: United Nations Environment Programme. Available at : www.unep.org/PDF/ PressReleases/GlobalMercuryAssessment2013.pdf

74. UNEP (2013a) *Mercury: time to act.* Nairobi: United Nations Environment Programme Available at: www.unep.org/PDF/PressReleases/Mercury_TimeToAct.pdf
75. Bellanger, M., Pichery, C., Aerts, D., Berglund, M., Castano, A., Cejchanova, M., Crettaz, P., Davidson, F., Esteban, M., Fischer, M.E., Gurzau, A.E., Halzlova, K., Katsonouri, A., Knudsen, L.E., Kolossa-Gehring, M., Koppen, G., Ligocka, D., Miklavcic, A., Reis, M.F., Rudnai, P., Tratnik, J.S., Weihe, P., Budtz-Jorgensen, E., Grandjean, P. (2013) Economic benefits of methylmercury exposure control in Europe: Monetary value of neurotoxicity prevention. *Environmental Health.*12: 3.
76. Greenwood, M, R. (1985) Methylmercury poisoning in Iraq. An epidemiological study of the 1971-1972 outbreak. *Journal of Applied Toxicology.*5(3):148-59.
77. Iustina Popescu, Rodica Stănescu, Mattia Biasioli, Franco Ajmone Marsan, Ionel Constantinescu, Assessing Human Risks Through *CSOIL* Exposure Model For A Soil Contamination Associated To Heavy Metals, U.P.B. Sci. Bull., Series B, Vol. 75, Iss. 1, 2013 ISSN 1454-2331,
78. Vern Grubinger and Don Ross, University of Vermont, Interpreting The Results Of Soil Tests For Heavy Metals, Sept. 15, 2011,
79. Directive 2004/35/CE– regarding the soil protection,
80. Directive proposal establishing a framework for the protection of soil - COM (2006) 232,
81. Romanian law - Order 756/1997 for the approval of the Regulation on the assessment of environmental pollution,
82. Health monitoring of the population in Baia Mare city in relation with exposure to lead, Public Health Department, Baia Mare, Maramureș 2008,
83. Health monitoring of the population in Baia Mare city in relation with exposure to lead, Public Health Department, Baia Mare, Maramureș, 2009,
84. Anca Maria Moldoveanu, Andra Neamțu, Daniela Rusea, Sima Carmen, Mariana Vlad, Biological screening of the population lead exposure, www.insp.gov.ro, 2012.

Citations

CHAPTER 1

G. U. Chibuike and S. C. Obiora, "Heavy Metal Polluted Soils: Effect on Plants and Bioremediation Methods," Applied and Environmental Soil Science, vol. 2014, Article ID 752708, 12 pages, 2014. doi:10.1155/2014/752708.

CHAPTER 2

Melanie Mehes-Smith, Kabwe Nkongolo and Ewa Cholewa (2013). Coping Mechanisms of Plants to Metal Contaminated Soil, Environmental Change and Sustainability, Dr. Steven Silvern (Ed.), ISBN: 978-953-51-1094-1, InTech, DOI: 10.5772/55124.

CHAPTER 3

Bieby Voijant Tangahu, Siti Rozaimah Sheikh Abdullah, Hassan Basri, Mushrifah Idris, Nurina Anuar, and Muhammad Mukhlisin, "A Review on Heavy Metals (As, Pb, and Hg) Uptake by Plants through Phytoremediation," International Journal of Chemical Engineering, vol. 2011, Article ID 939161, 31 pages, 2011. doi:10.1155/2011/939161.

CHAPTER 4

Michaela Meyer, Roland Pesch, Winfried Schröder, Eiliv Steinnes, and Hilde Thelle Uggerud, Spatial Patterns and Temporal Trends of Heavy Metal Concentrations in Moss and Surface Soil Specimens Collected in Norway between 1990 and 2010, doi:10.1186/s12302-014-0027-0.

CHAPTER 5

Baoling Wang, Yu Wang, and Weidong Wang, Retention and Mitigation of Metals in Sediment, Soil, Water, and Plant of A Newly Constructed Root-Channel Wetland (China) from Slightly Polluted Source Water, doi:10.1186/2193-1801-3-326.

CHAPTER 6

Alpana Kulhari, Arun Sheorayan Somvir Bajar Susheel Sarkar, Ashok Chaudhury, andRajwant K Kalia, Investigation of heavy metals in frequently utilized medicinal plants collected from environmentally diverse locations of north western India, doi:10.1186/2193-1801-2-676.

CHAPTER 7

Sayyed Mohammad Javad Mirzaei, Manouchehr Heidarpour, Sayyed Hassan Tabatabaei, Payam Najafi, and Seyyed Ebrahim Hashemi, Immobilization of leachate's heavy metals using soil-zeolite column, doi: 10.1186/2251-7715-2-20.

CHAPTER 8

Anca Maria Moldoveanu (2014). Assessment of Soil Pollution with Heavy Metals in Romania, Environmental Risk Assessment of Soil Contamination, Dr. Maria C. Hernandez Soriano (Ed.), ISBN: 978-953-51-1235-8, InTech, DOI: 10.5772/57595.

Index

D

E

F

G

H

I

L

M

N

O

P